An introduction to

Biochemical Aspects of the
Adrenal Cortex

An introduction to

Biochemical Aspects of the Adrenal Cortex

JOHN S. JENKINS

M.A., M.D.(Cantab). M.R.C.P.

Physician and Endocrinologist,
St. George's Hospital, London

EDWARD ARNOLD (Publishers) LTD
LONDON

© John S. Jenkins, 1968

First published 1968

SBN: 7131 4142 5

Printed in Great Britain by
Page Bros. (Norwich) Ltd., Norwich.

PREFACE

During the last two decades interest in the adrenocortical hormones has extended outside the realms of endocrinology to involve almost every branch of medicine. The present volume is intended to introduce the biochemical aspects of adrenocortical function to those physicians and pathologists who are without specialist knowledge of the subject. The first two sections are devoted to a consideration of normal and abnormal adrenocortical secretion and the third section provides a survey of the laboratory techniques which are currently in use for the investigation of adrenocortical problems. Particular emphasis has been placed on those methods which are within the capabilities of an ordinary hospital laboratory.

In a volume of this size, references to the enormous amount of literature relating to the adrenal steroids are of necessity highly selective, and cannot do justice to the many original workers who have provided the real advances in this field. Sometimes the price of brevity is dogmatism, but it is hoped that use of the review articles and other works quoted will help to dispel any such failings.

I wish to express my thanks to colleagues at St George's Hospital for sustaining my interest in the adrenal cortex over the years, and to Mrs Betty Simpson who had the arduous task of typing the manuscript. I am indebted to Messrs E. and S. Livingstone Ltd, for permission to publish Fig. 4.5, and Pitman Medical Publishing Co., for Fig. 5.3.

J.S.J.

London, 1968

CONTENTS

PART I

Normal Adrenocortical Function

1

THE ADRENOCORTICAL HORMONES

Biologically active extracts of adrenal cortical tissue were prepared as long ago as 1927 by Hartman, but it was during the years 1935–42 that the isolation and crystallization of adrenal steroids was achieved by the efforts of several groups of workers in Switzerland and the U.S.A., who used enormous amounts of bovine adrenal glands to produce a few milligrams of crystalline material. During this period, cortisone, corticosterone, and cortisol were characterized, and the first synthesis of a corticosteroid, deoxycorticosterone, was achieved by Reichstein in Basle. The adrenal residue after crystalline substances had been removed was also known to be active, but it was not until 1952 that Simpson and Tait in London observed that this amorphous fraction contained a substance which was a potent regulator of sodium and potassium concentrations, and is now known as aldosterone. Kendall at the Mayo Clinic had previously been successful in isolating cortisone, and in 1949 his clinical colleagues Hench and others reported the dramatic results of treatment of a non-endocrine disease, rheumatoid arthritis, with cortisone. These findings provided a sudden tremendous impetus to the study of the adrenal cortex and the chemistry of its secretion.

Although well over 40 steroids have now been recovered from adrenal extracts, some are present in only trace amounts, some are intermediate products only, and some are probably artifacts of the extraction process.

THE ADRENAL CORTICAL SECRETION

From the biological point of view particular interest is attached to the steroids present in the adrenal venous blood, especially those in greater concentration than in the peripheral blood.

Corticosteroids

The nature of the adrenal cortical secretion varies according to the animal species, but studies on the adrenal blood of man show that cortisol (hydrocortisone) and aldosterone are the most important, and, in fact, patients

1

with adrenal insufficiency can be kept in good health by the regular administration of these two hormones or their equivalents. The human gland also produces some corticosterone, a steroid which is the predominant one in some animals, including the rat, mouse, and rabbit. The normal total daily secretion of cortisol is approximately 20 mg, corticosterone 2 mg, and aldosterone 0·1 mg, and the concentrations of these steroids in the peripheral blood are of the order of 10 μg, 1μg, and 0·01 μg per 100 ml. of plasma, respectively. These amounts vary considerably, however, according to the influence of the mechanisms controlling the secretion of the hormones. It is of interest that cortisone, the first of the corticosteroids to be used in treatment of nonendocrine disease, is present in only trace amounts in adrenal blood and is generally considered to be a metabolite of cortisol.

Corticosteroids in the blood are discussed in detail by Dixon *et al.* (1967).

Adrenal androgens

In addition to these corticosteroids which have 21 carbon atoms in their chemical structure (see Chapter 2) the adrenal also produces androgenic steroids which have 19 carbon atoms (Figs. 2.11–13). The most important examples of this class of steroid in the adrenal venous blood are Δ^4- androstenedione, (androst-4-ene-3,17,-dione), 11β-hydroxy-Δ^4-androstenedione, and dehydroepiandrosterone. The latter was formerly believed to be exclusively adrenal in origin but it is now known to be secreted by the gonads also, under some conditions. The androgenic activity of the adrenal C_{19} steroids is low compared with that of testosterone but they assume importance in certain disorders of adrenal steroid biosynthesis (see Chapter 7).

Adrenal oestrogens

It has been believed for some years that the normal adrenal cortex secretes oestrogens, which are steroids with 18 carbon atoms, and the treatment of metastasizing breast cancer by bilateral adrenalectomy is founded on this possibility. Biochemical evidence of oestrogen secretion has not been easy to obtain, mainly owing to the very small quantities secreted, but an increase in oestrogen excretion has been found in the urine of ovariectomized women whose adrenals have been stimulated with adrenocorticotrophic hormone (ACTH). The possibility remains however, that adrenal androgen is converted to oestrogen in extra-adrenal tissues such as the liver.

Protein-binding of adrenal steroids

When cortisol enters the circulation it is not carried in the blood in simple solution but is largely bound to protein. Two proteins are involved. One is an α-globulin known as corticosteroid-binding globulin or transcortin, which has a high affinity for cortisol but a low capacity. The other protein which binds cortisol is albumin and has a very high capacity but its affinity is less than that of corticosteroid-binding globulin. Cortisol is normally 94% protein-bound and there is good evidence that only the small unbound fraction is actually available for biological activity. In pregnancy it is well established that during the last trimester cortisol levels in the blood may rise into a range equal to or even greater than that seen in Cushing's syndrome, and yet there is no clinical evidence of a hyper-adrenal state. The explanation lies in the increase in corticosteroid-binding globulin which occurs during pregnancy, so that more cortisol is protein bound but the unbound fraction is generally considered to be normal. Some authors, however (Plager *et al.* 1964), claim that the unbound fraction is also greater during pregnancy, in which case it is difficult to account for the apparently normal adrenal state of the patient. The administration of oestrogens, likewise, increases the amount of corticosteroid-binding globulin with a resulting increase in total plasma cortisol, and abnormally high levels are found in patients taking oral contraceptives. Decreased corticosteroid-binding globulin is found in new-born infants in whom the unbound fraction occupies a far greater proportion of the total than in the adult. Decreased cortisol-binding is also found in conditions associated with disorders of serum proteins such as multiple myelomatosis and nephrosis. Corticosterone is also bound to corticosteroid-binding globulin, although the affinity is rather less for this steroid than for cortisol. Aldosterone is protein-bound to a much less extent than either cortisol or corticosterone, and is mainly associated with albumin. Protein-binding of steroids is reviewed by Mills (1962).

The structure of the adrenal cortex and its relation to steroid secretion

The treatment of some cases of carcinoma of the breast by bilateral adrenalectomy has provided much information about the morphology and steroid content of the normal human adrenal gland during life. The normal adrenals each weigh about 4 grams and are rather smaller than was formerly believed from studying only post-mortem material. Histologically, the adrenal cortex is divided into three layers; an outermost zona glomerulosa, which lies immediately below the capsule; an intermediate zona fasciculata composed of long columns of cells filled with lipid material; an innermost zona reticularis composed of compact cells poor in lipid.

In man, unlike the rat, the zona glomerulosa is not very prominent, but studies on animal tissue carried out *in vitro* show that this layer is responsible for the production of aldosterone, and there is evidence that this also occurs in the human gland (Stachenko & Giroud, 1962). Hypophysectomy leads to rapid atrophy of the two inner zones while the zona glomerulosa remains intact for a considerable time, indicating that factors other than the anterior pituitary are important in the control of aldosterone secretion. According to some authors the zona fasciculata is responsible for the synthesis of cortisol and corticosterone, and the zona reticularis synthesizes androgens and oestrogens. Extensive studies, however, by Symington (1962) provide good evidence that the zona reticularis and fasciculata act together to form all the adrenal steroids other than aldosterone. Symington's group believe that the action of ACTH is directed especially at the cells of the zona fasciculata, which contain a store of steroid precursors, although once synthesis of steroids has taken place they are rapidly released from the gland. Studies on the vasculature of the adrenal gland by Dobbie & Symington (1966) indicate that the muscular nature of the veins may be important in determining the release of hormones following stress or ACTH.

The foetal adrenal cortex

The adrenal of the human foetus is, relative to its body size, much larger than that of the adult, and this increase is accounted for mainly by a distinct innermost layer known as the foetal zone. This layer rapidly involutes within a few days after birth, but it has been shown from studies *in vitro* that from the 22nd week of intra-uterine life onwards, the foetal cortex can synthesize cortisol and corticosterone. It does not, however, synthesize aldosterone, so that the foetal zone appears to function in a similar manner to the zona fasciculata and reticularis of the adult organ. The absence of a large foetal zone in anencephalic infants indicates that the foetal pituitary controls the size of the adrenal cortex (Jost, 1966). There is probably, however, some dissociation of the different components of adrenal secretion in these infants, since the small adrenal may produce normal amounts of cortisol but little or no androgen.

REFERENCES

DIXON, P. F., BOOTH, M. and BUTLER, J. (1967). In *Hormones in the Blood*, Vol. 2. pp. 305–389. Ed. by C. H. GRAY and A. L. BACHARACH. Academic Press, London and New York.

DOBBIE, J. W. and SYMINGTON, T. (1966). *J. Endocr.*, **33,** 479–489.

JOST, A. (1966). *Recent Prog. Hormone Res.*, **22,** 541–574.

MILLS, I. (1962). *Brit. Med. Bull.* **18,** 127–133.

PLAGER, J. E., SCHMIDT, K. G. and STAUBITZ, W. J. (1964). *J. clin. Invest.*, **43,** 1066–1072.

STACHENKO, J. and GIROUD, C. J. P. (1962). In *The Human Adrenal Cortex*, pp. 30–43, Ed. by A. R. CURRIE, T. SYMINGTON, and J. K. GRANT. Livingstone, Edinburgh.

SYMINGTON, T. (1962). In *The Human Adrenal Cortex*, pp. 3–20. Ed. by A. R. CURRIE, T. SYMINGTON, and J. K. GRANT, Livingstone, Edinburgh.

2

THE CHEMISTRY OF THE
ADRENOCORTICAL HORMONES

The adrenocortical hormones are steroids, a class of compounds which is widely found in nature, and includes such diverse substances as cholesterol, the sex hormones, and the cardiac aglycones of digitalis. The basic chemical structure of the steroid hormones with the conventional numbering of the carbon atoms is shown in Fig. 2.1. The essential features are three fused cyclohexane rings A, B, C, and a cyclopentane D, drawn for convenience,

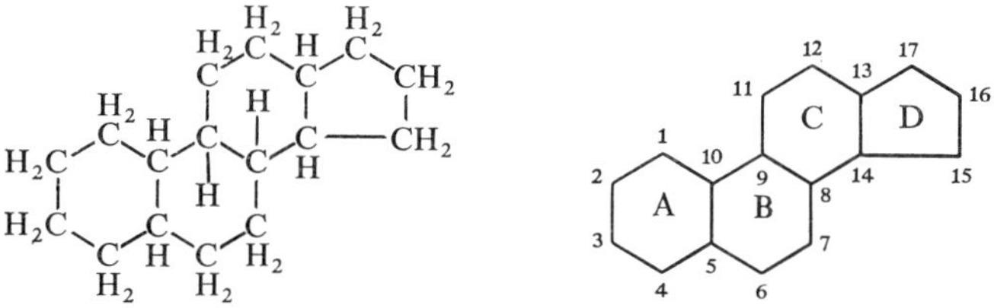

(a) Full structure   (b) Conventional formula

FIG. 2.1. Basic steroid nucleus.

as three hexagons and a pentagon. Two methyl groups project above the general plane of the molecule at positions 13 and 10, and the carbon atoms of these methyl groups are numbered 18 and 19 respectively. The steroids which are characteristic of the adrenal cortex have 21 carbon atoms (C_{21} compounds) and belong to what is termed the "pregnane" series (Fig. 2.2.) in which there is a two-carbon side chain attached to carbon 17.

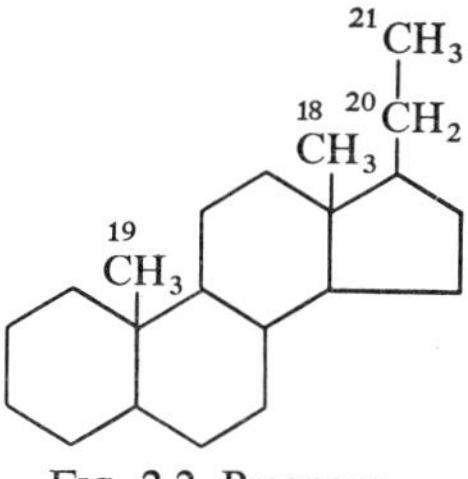

FIG. 2.2. Pregnane.

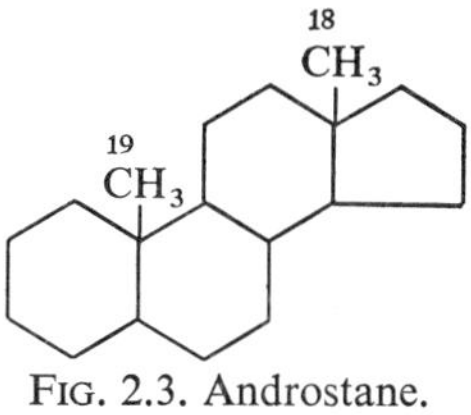

FIG. 2.3. Androstane.

5

Androgenic steroids secreted by the adrenal cortex and by the testes have 19 carbon atoms (C_{19} compounds) and belong to the general series called androstane (Fig. 2.3). The addition of side groups at various points of the skeleton determines the particular biological properties of the individual steroids.

Stereochemistry

In addition to the nature and position of the side groups, their spatial relationships are also extremely important so that a knowledge of the stereochemistry of the steroids is necessary when attempting to relate chemical structure with physiological action.

The side groups may project either in the general plane of the rather flat molecule (equatorial) or perpendicular to the plane (axial). The axial groups may project above the plane in the same direction as the methyl groups at carbons 10 and 13, and are then designated β, the bond being drawn as a solid line. If they project below the plane they are called α, drawn as a dashed line. By the rotation or twisting of bonds the six carbon atoms of cyclohexane rings may take up different positions in space known as conformations, which are commonly described as "chair" or "boat" forms (Fig. 2.4).

"Chair"   "Boat"

FIG. 2.4. Conformations of cyclohexane.

The "chair" is the most stable arrangement and is the usual one to be found in the steroids. The rings may also take up one of two possible spatial arrangements at their junctions (Fig. 2.5). Those adrenal steroids which have saturated A and B rings can have the axial hydrogen at carbon 5 either in the α- or β-position. If the hydrogen atom is in the α-position, rings A and B assume a form known as the "*trans*" configuration, and if the hydrogen is in

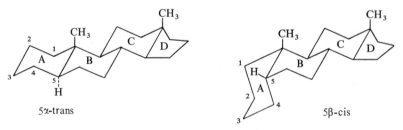

5α-trans   5β-cis

FIG. 2.5. Configurations at ring junctions.

the 5 β-position the two rings assume the position known as the *"cis"* form. In the adrenal steroids the other ring junctions are invariably *trans*. Most of the metabolites of cortisol found in human urine are in the 5β-,*cis*, form, whereas in the case of the C_{19} compounds the *cis* and *trans* isomers are more equally represented.

The nomenclature of adrenal steroids

It is convenient to refer to the adrenal steroids, as far as possible, by their common or "trivial" names, e.g. cortisol, corticosterone, aldosterone. When, however, it is necessary to describe fully the chemical structure of a steroid, the "systematic" name is used, and this is formulated according to certain accepted rules.

The side groups which characterize a particular steroid are indicated by prefixes, or suffixes, attached to pregnane or androstane. The position of a group is given by the number of the carbon atom to which it is attached and where appropriate, its spatial relationship to the molecule is indicated by α or β. Hydroxyl groups are given as prefixes unless they are the only substituents when they are given as suffixes. Double bonds and carbonyl groups are given as suffixes, in that order. Prefixes and suffixes used are:

	Prefix	Suffix
Hydroxyl, —OH	hydroxy-	-ol
Double bond C=C	Δ, (not now correct in systematic names)	-ene
Carbonyl, C=O	oxo- (formerly keto)	-one

The position of a double bond is indicated by the lowest number of the two carbon atoms if they are consecutive, e.g. Δ^4, or 4-ene, refer to a double bond between carbons 4 and 5. If the carbon numbers are not consecutive, both are given, e.g. 9,11-ene. The following examples illustrate the use of systematic names:

Trivial name	Systematic name	Formula
Cortisol (Hydrocortisone is unofficial)	11β, 17α, 21-trihydroxy-pregn-4-ene-3,20-dione	Fig. 2.6.
Cortisone	17α,21-dihydroxy-pregn-4-ene-3.11,20-trione	Fig. 2.7.
Dehydroepiandrosterone	3β-hydroxy-androst-5-ene-17-one	Fig. 2.13.

When ring A is saturated the spatial position of the hydrogen atoms at carbon 5 is indicated, e.g. tetrahydrocortisol is 3α,11β,17α,21-tetrahydroxy-5β-pregnane-20-one (Fig. 4.1).

B

Trivial names are also modified by the addition of certain prefixes as follows:

allo-	sometimes used to indicate 5α-compounds.
epi-	inversion of the usual position of a group, e.g. epiandrosterone in which there is 3β-hydroxy instead of 3α-hydroxy in androsterone.
dehydro-	loss of 2 hydrogen atoms from adjacent carbon atoms to form a double bond, or from a —CHOH group to form a carbonyl (keto-) group.
deoxy- (desoxy- in U.S.A.)	replacement of a hydroxyl group by hydrogen, e.g. 11-deoxycorticosterone (Fig. 2.9).
dihydro-	addition of 2 hydrogen atoms to saturate a double bond or to reduce a carbonyl group to form a secondary alcohol.
tetrahydro-	addition of 4 hydrogen atoms, usually to saturate a double bond together with the reduction of a nearby carbonyl group to form a secondary alcohol, e.g. tetrahydrocortisol.
oxy-	denotes the presence of an oxygen atom either as hydroxyl or carbonyl. It is to be distinguished from the prefix oxo-, which refers only to a carbonyl group, e.g. 11-oxy-17-oxo-steroids.

THE RELATIONSHIP BETWEEN CHEMICAL STRUCTURE AND BIOLOGICAL ACTIVITY

Naturally occurring steroids

The predominant steroid found in human adrenocortical secretion is cortisol, the formula of which is shown in Fig. 2.6. The double bond in ring A between carbons 4 and 5 and the ketone group at carbon 3 are

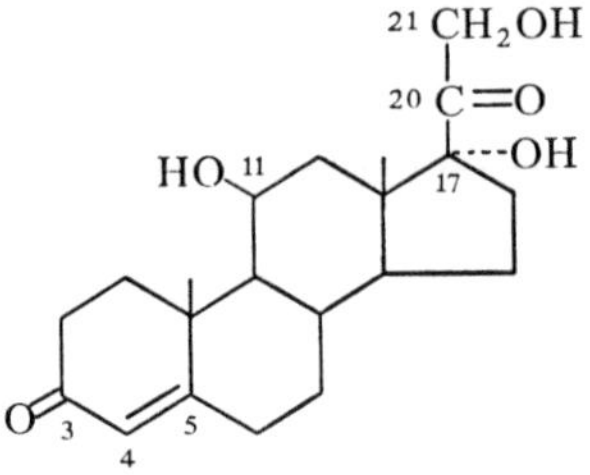

FIG. 2.6. Cortisol. 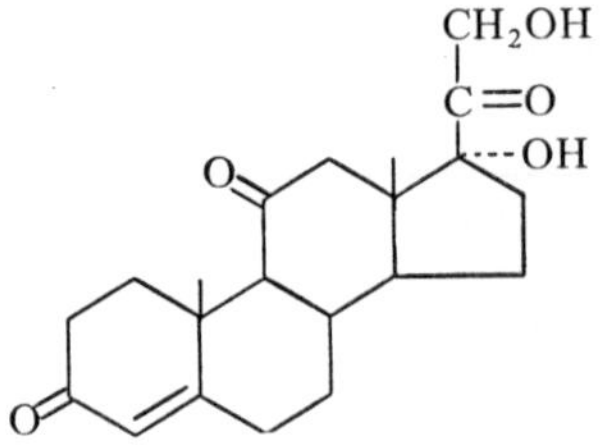   FIG. 2.7. Cortisone.

necessary for the biological activity of many steroids. The hydroxyl group at carbon 11 in the β-position is essential for the glucocorticoid activity of cortisol (see Chapter 6) and with very rare exceptions is specifically adrenal in origin. The hydroxyl at carbon 21 and the ketone group at carbon 20 are also necessary for full glucocorticoid activity. Absence of the 17α-hydroxyl group is accompanied by a decrease in glucocorticoid activity but an increase in mineralocorticoid properties. The resultant steroid is corticosterone (Fig. 2.8), a minor component of human adrenal secretion but the predominant steroid in the rat and some other animals.

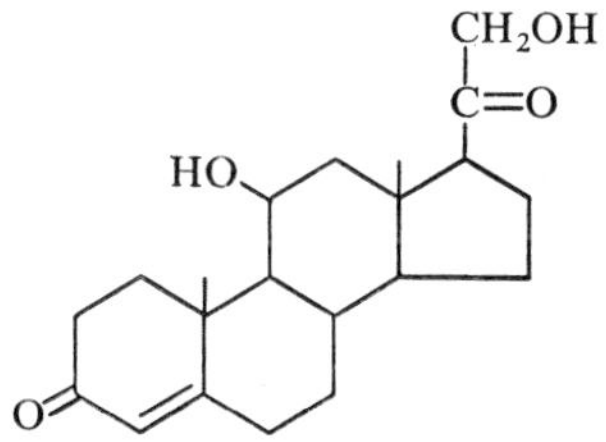

FIG. 2.8. Corticosterone.

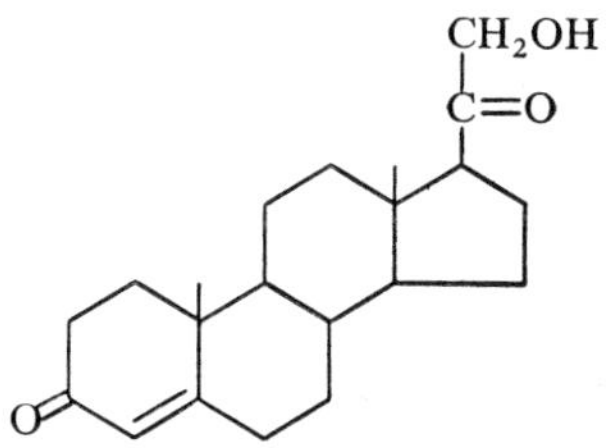

FIG. 2.9. Deoxycorticosterone.

It is of interest that cortisone (Fig. 2.7), perhaps the best known of the adrenal steroids because it was the first glucocorticoid to be used therapeutically, is itself biologically inactive, owing to the presence of a ketone group at carbon 11. Cortisone becomes active only after its 11-ketone has been reduced to the 11β-hydroxyl in the liver to form cortisol. Aldosterone, the other important constituent of adrenal cortical secretion, is an example of a mineralocorticoid (see Chapter 6) and its chemical structure is shown in Fig. 2.10. It lacks the 17α-hydroxyl group of cortisol, but the most striking characteristic is the presence of an aldehyde group at carbon 18. The elucidation of this unusual structure was carried out by Simpson and Tait, in

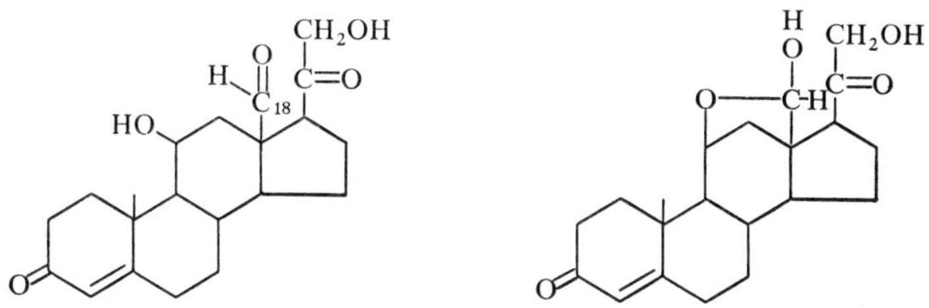

(a) Open form

(b) Hemiacetal

FIG. 2.10. Aldosterone.

collaboration with Reichstein and Wettstein in 1954, only two years after the original discovery of the hormone. The hemiacetal formed by association of the 11β-hydroxyl and the 18-aldehyde is the more stable state (Fig. 2.10b). Dextro- and laevo-rotatory forms are also possible but only d-aldosterone, the naturally occurring form, is biologically active. The artificial synthesis of the hormone results in a racemic mixture so that only half of this is therapeutically effective. Although the 11β-hydroxyl is present in aldosterone, it is not essential for mineralocorticoid activity since it is absent in deoxycorticosterone (Fig. 2.9), a compound which was used therapeutically as a mineralocorticoid for many years before the discovery of aldosterone. Deoxycorticosterone is, however, about 30 times less active than aldosterone.

The naturally occurring adrenal androgens are shown in Figs. 2.11, 2.12, and 2.13. They are all relatively weak in their androgenic effects and the

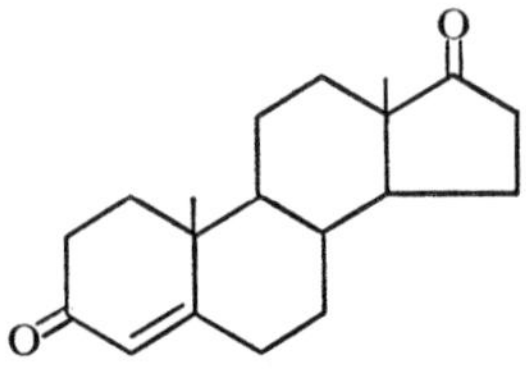

Androst-4-ene-3,17-dione	11β-hydroxyandrost-4-ene-3,17-dione	Dehydroepiandrosterone
FIG. 2.11.	FIG. 2.12.	FIG. 2.13.
	Adrenal Androgens	

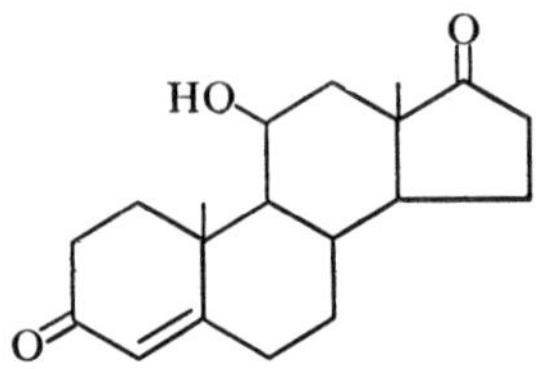

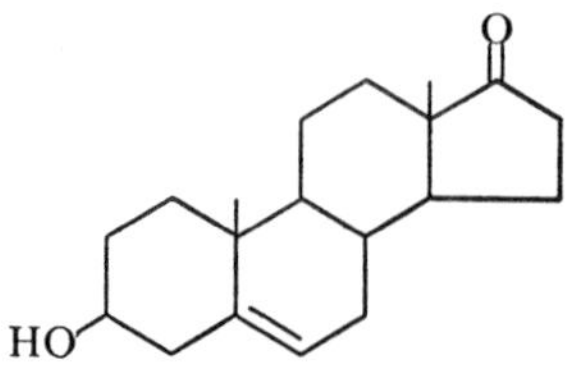

introduction of a 11β-hydroxyl group makes 11β-hydroxyandrost-4-ene-3,-17-dione the weakest of all. Testosterone (Fig. 2.14) the very potent gonadal androgen, has not been identified in the normal adrenal but has been isolated from some adrenal tumours.

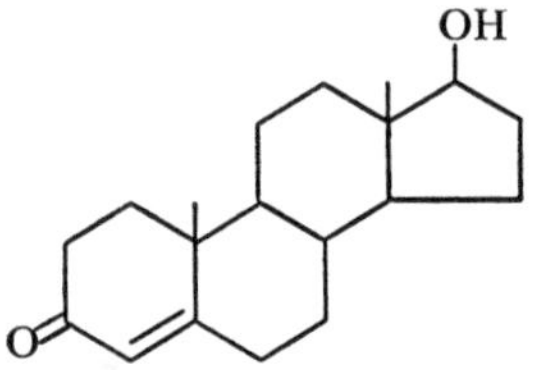

Testosterone

FIG. 2.14.

Any chemical alteration in the side groups of the adrenal steroids or a change in their spatial arrangement usually results in either a decrease or complete loss of biological activity.

Synthetic derivatives of cortisol

The initial enthusiasm which greeted the therapeutic successes of cortisol and cortisone in a wide variety of diseases was later tempered by recognition of the considerable side effects which often accompanied their use. There was, as a result, a great stimulus to the steroid chemists to produce derivatives of the natural hormone, cortisol, which would have even greater therapeutic activity but with less undesirable side effects. Increased potency has been achieved, often to a striking degree, but removal of unwanted actions has not been possible except in one important instance, the dissociation of glucocorticoid from mineralocorticoid activity. The first of these synthetic analogues were prednisolone (Fig. 2.15) and the corresponding 11-oxo-(keto) steroid prednisone (Fig. 2.16). It can be seen that the only difference from

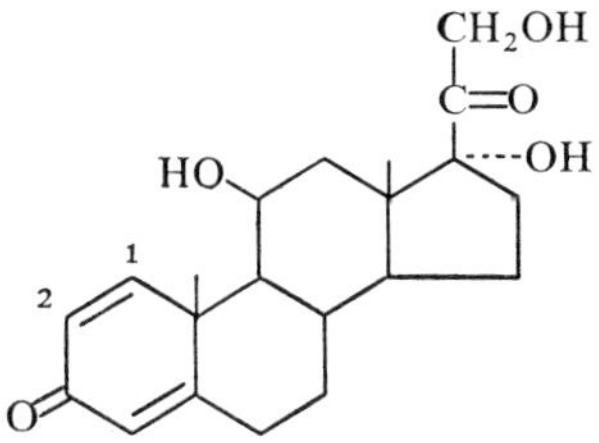

FIG. 2.15. Prednisolone.

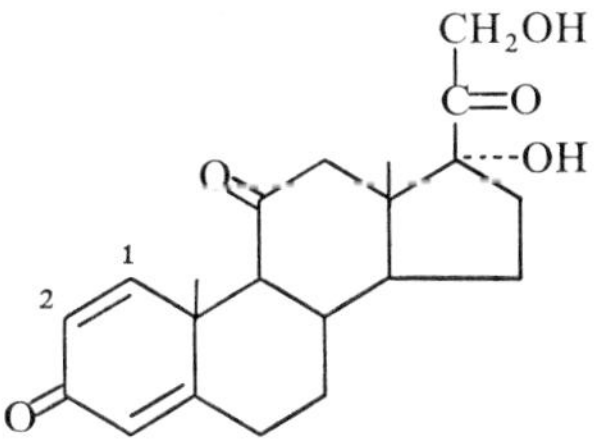

FIG. 2.16. Prednisone.

cortisol and cortisone is the presence of a second double bond in ring A between carbons 1 and 2. This modification makes ring A more inaccessible to the enzymes which normally inactivate cortisol by reduction of the 3-ketone and 4,5 double bond. The result is that the glucocorticoid activity of prednisolone is five times that of cortisol. The introduction of a methyl group at carbon 6 (Fig. 2.18) also has the effect of protecting the ring A structure. Some of the most interesting modifications of cortisol have resulted from the introduction of a halogen in the α-position at carbon 9, and of all the halogens fluorine has by far the most effect (Fried & Borman, 1958). Glucocorticoid activity is considerably increased, but mineralocorticoid activity is increased to the spectacular extent of 300 times that of the parent steroid, so that 9α-fluorocortisol (Fig. 2.17) is equal in this respect to the powerful

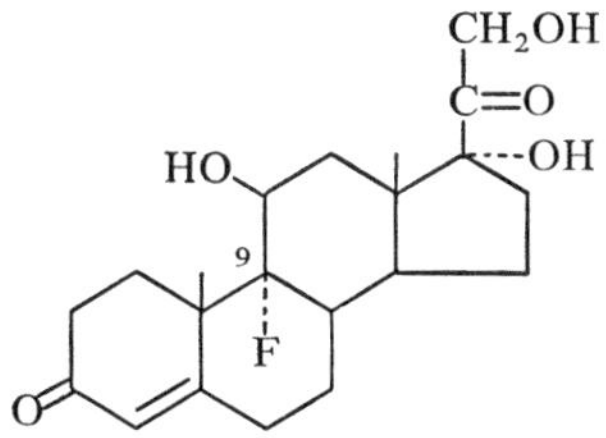

FIG. 2.17. 9α-fluorocortisol.

natural mineralocorticoid aldosterone. Since fluorocortisol is easier to synthesize than aldosterone and is fully effective when given orally, it is used in place of the natural hormone in the treatment of Addison's disease. The action of the 9α-fluorine atom is complex and has received much attention by Fried and Borman. It is significant that the most active halogen, fluorine, is also the most electro-negative, and the effect of this property on the neighbouring 11β-hydroxyl group is to increase the acidity and hence the hydrogen-donating tendency of this biologically important group.

A further effect of the substitution at the 9α-position may be to hinder changes in the ring A structure, but more important, according to Bush & Mahesh (1964), is the prevention of oxidation of the 11β-hydroxyl group.

The introduction of an α-hydroxyl at carbon 16 has the effect of decreasing the inactivation of the side chain attached to carbon 17, especially the reduction of the ketone group at carbon 20, but the effect on 9α-fluoro-cortisol is to remove completely the mineralocorticoid activity. If the double bond at carbon 1 and 2 of prednisolone is also present the result is triam-cinolone (Fig. 2.19) which is a glucocorticoid with about seven times the

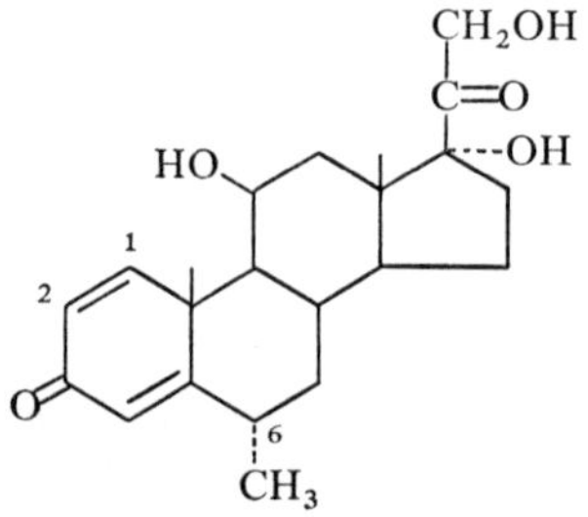

FIG. 2.18. 6α-methylprednisolone.

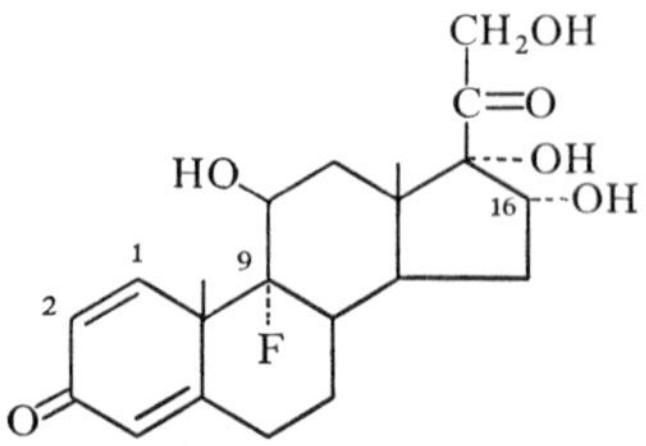

FIG. 2.19. Triamcinolone (16α-hydroxy, 9α-fluoroprednisolone).

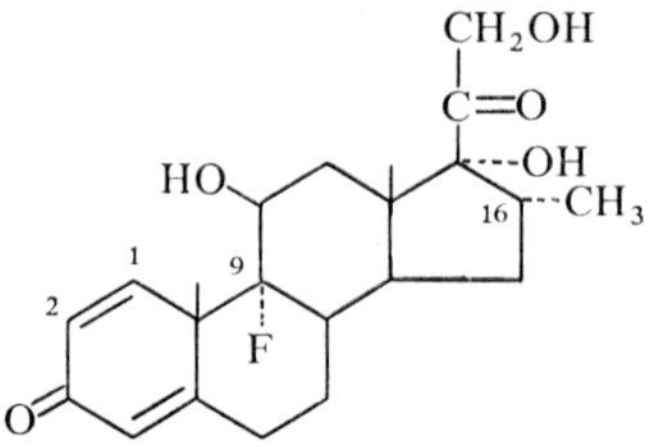

FIG. 2.20. Dexamethasone (16α-methyl, 9α-fluoroprednisolone).

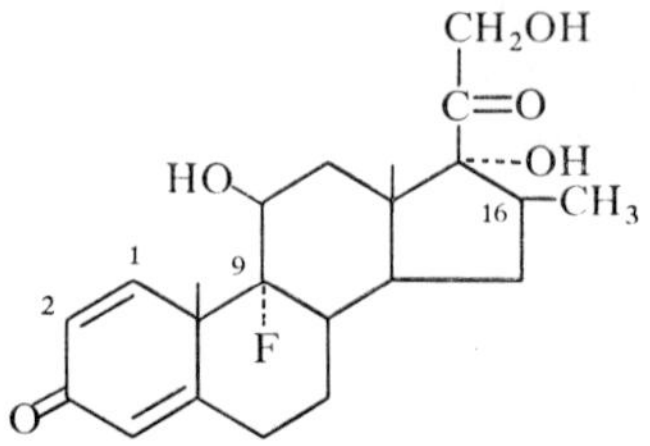

FIG. 2.21. Betamethasone (16β-methyl, 9α-fluoroprednisolone).

activity of cortisol. An even more impressive increase in glucocorticoid activity has been achieved by the synthesis of dexamethasone (Fig. 2.20) which differs from triamcinolone in having an α-methyl group at carbon 16. By reason of the substitutions in the cortisol structure which comprise dexamethasone, reduction of the 3-ketone and the 4,5 double bond are hindered, oxidation of the 11β-hydroxyl is prevented and its reactivity increased, and the reduction of the 20-ketone is diminished by the 16α-methyl group. The result is a compound with 30 times the glucocorticoid activity of cortisol but with no mineralocorticoid action whatsoever. Betamethasone (Fig. 2.21) is closely similar to dexamethasone, having a 16β- instead of a 16α-methyl group, but has somewhat less activity.

It can be seen that many of the modifications introduced into the cortisol molecule have the effect of reducing the inactivating processes which normally occur during the metabolism of cortisol, or they actually increase the reactivity of the important 11β-hydroxyl group. There are, however, other factors responsible for the increased potency of the synthetic cortisol analogues. Changes in solubility may lead to a greater rate of absorption but of special importance is the degree of binding of the steroid by the corticosteroid-binding proteins in the blood. Prednisolone is bound to about the same degree as cortisol itself but steroids with substitutions at the 9α- and the 16 positions are bound to a much less extent (Florini & Buyske, 1961) so that their biological effectiveness is correspondingly increased.

The exact manner in which the structure of the adrenocortical steroids determines their physiological action is not known since the mode of action of steroids at the cell level is not understood with any certainty. Bush (1962) has suggested that in the case of the glucocorticoids there is a close attachment of the flat β-surface of rings C and D to a cell "receptor". The importance of the 11β-hydroxyl points to a hydrogen bond between this group and a proton-accepting group in the receptor, perhaps aided by attachment of the side chain containing the 21-hydroxyl and 20-ketone groups. Such considerations are of interest but must still be regarded as largely speculative.

A good introduction to the chemistry of the whole class of steroids is to be found in the monograph of Klyne (1957).

REFERENCES

BUSH, I. E. (1962). In *The Human Adrenal Cortex.*, pp. 138–171. Ed. by A. R. CURRIE, T. SYMINGTON, and J. K. GRANT. Livingstone, Edinburgh.
BUSH, I. E. and MAHESH, V. B. (1964). *Biochem. J.* **93**, 236–255.
FLORINI, J. R. and BUYSKE, D. A. (1961). *J. Biol. Chem.* **236**, 247–251.
FRIED, J. and BORMAN, A. (1958). *Vitam. and Horm.* **16**, 303–374.
KLYNE, W. (1957). *The Chemistry of the Steroids.* Methuen, London.

3

THE BIOSYNTHESIS OF THE ADRENAL STEROIDS

The results of experiments carried out both *in vivo* and *in vitro* have shown that cholesterol is the main intermediate in the synthesis of the adrenal steroids. Adrenal tissue can also convert the simpler molecule, acetate, to steroids, and by the use of acetate with several carbon atoms radioactively labelled it has been shown that cholesterol is first formed. The synthesis of cholesterol from acetate has been fully described by Popjak & Cornforth (1960). The possibility that cholesterol is not an obligatory intermediate has been considered, and there is evidence that other sterols may act as precursors since Werbin *et al.* (1960) reported that β-sitosterol could be converted to cortisol by the guinea pig. While the adrenal itself can undoubtedly synthesize cholesterol, some investigators believe that the main source for steroid synthesis *in vivo* is cholesterol derived from the plasma. Within the human adrenal, about 20% of the cholesterol is unesterified and about 80% is esterified with various fatty acids. By the use of radioactively labelled material it seems, at least in vitro, that the free, unesterified form, is utilized for steroid synthesis (Jenkins, unpublished data). The first step in the synthesis of steroids from cholesterol is the formation of 20α,22-dihydroxy cholesterol, following which the side chain of this C_{27} compound is split off enzymatically by a desmolase present in the mitochondria to form the C_{21} steroid pregnenolone (Fig. 3.1), together with isocaproic aldehyde. Reduced nicotinamide adenine dinucleotide phosphate ($NADPH_2$) is an essential cofactor in this reaction.

Biosynthesis of Cortisol and Corticosterone (Fig. 3·1)

Pregnenolone is converted to progesterone by two enzyme systems found in the microsomes of the cell. One enzyme is a 3β-hydroxysteroid dehydrogenase which oxidizes the 3β-hydroxy group to a 3-oxo-group and requires nicotinamide adenine nucleotide (NAD) as a hydrogen-acceptor; the other is a Δ^5-3-oxosteroid isomerase which causes the transfer of the double bond from the 5, 6 position to the 4, 5 position. Progesterone is an important

15

hormone in its own right, and, as such, is synthesized by the corpus luteum of the ovary, but in the adrenal cortex it is a key intermediate compound. In the formation of cortisol and corticosterone a sequence of hydroxylations then takes place. Cortisol synthesis requires first the 17α-hydroxylation of progesterone, which is followed by 21-hydroxylation to give 11-deoxycortisol,

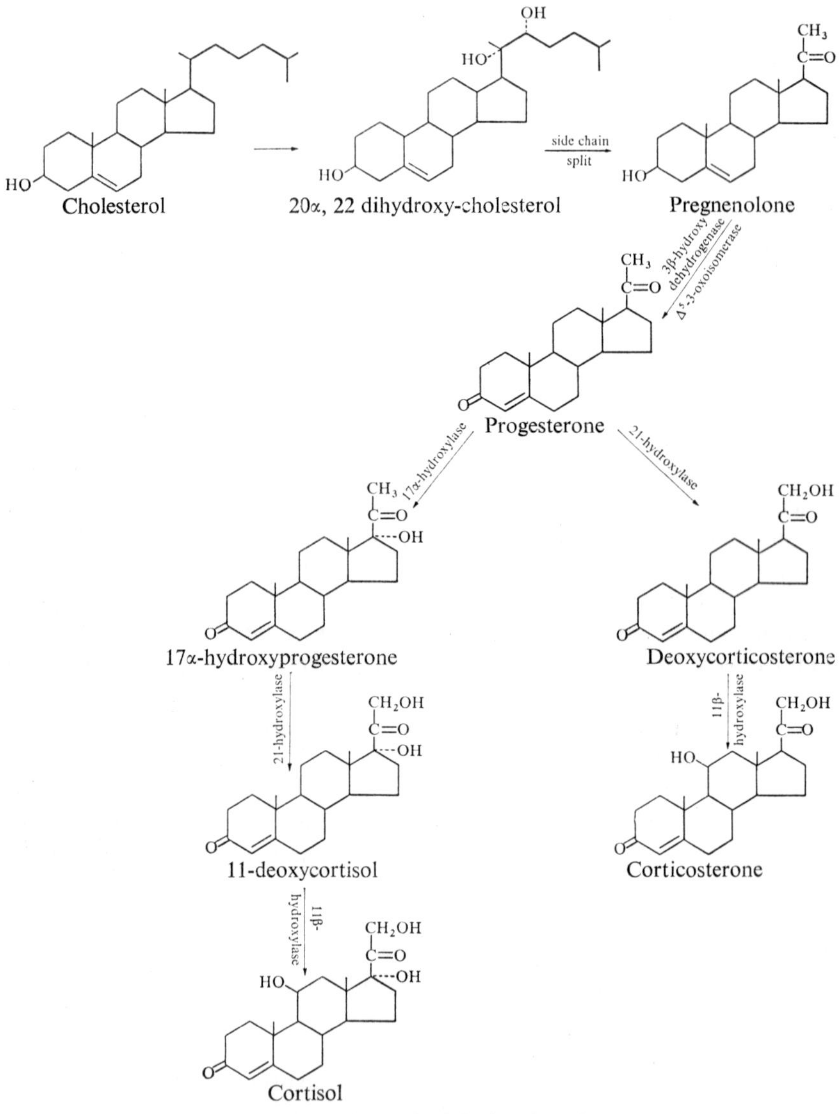

FIG. 3.1. The biosynthesis of cortisol and corticosterone.

and finally 11β-hydroxylation occurs. While a different order of hydroxylation is possible, the predominant sequence is shown in Fig. 3.1. If 17-hydroxylation does not take place, 21-hydroxylation of progesterone produces deoxycorticosterone which is followed by 11β-hydroxylation to yield corticosterone. The 17-hydroxylation of corticosterone to form cortisol does not appear to be possible. These three hydroxylase enzymes all require $NADPH_2$ as cofactor together with molecular oxygen, but they differ in their intracellular sites of origin. The 17α- and 21-hydroxylases are associated with the microsomes and the 17α-hydroxylase is found not only in the adrenal but also in the testis and ovary. The 11β-hydroxylase has been the most studied of the enzymes and, with very rare exceptions, is entirely adrenal in origin. It is associated with the mitochondria and probably consists of at least four protein constituents. Not all 11-deoxysteroids are equally effective as substrates for the enzyme, since 11-deoxycorticosterone and 11-deoxycortisol are hydroxylated to a much greater extent than androst-4-ene-3,17-dione. It has been suggested that there may be more than one 11β-hydroxylase system.

Biosynthesis of aldosterone (Fig. 3.2)

Incubation experiments with various substrates using the outermost layer of adrenal cortex suggest that the major pathway from cholesterol to aldosterone is first through the sequence of progesterone, deoxycorticosterone, and corticosterone. The 18-hydroxyl derivative of corticosterone is then probably formed since 18-hydroxycoticosterone has been isolated *in vitro* from incubates of zona glomerulosa. While corticosterone is formed in all zones of the adrenal cortex, the 18-hydroxylase enzyme is confined to the zona glomerulosa; no cortisol is formed by this layer owing to the absence of 17-hydroxylase. 18-hydroxycorticosterone is then presumably converted to the 18-aldehyde form to give aldosterone.

Biosynthesis of adrenal androgens (Fig. 3.3)

One pathway lies in the formation of progesterone and 17-hydroxyprogesterone, thence the side chain attached to carbon 17 is split off by a desmolase enzyme to yield androst-4-ene-3,17-dione. Some of this may then be hydroxylated at the 11β-position by the 11β-hydroxylase. Dehydroepiandrosterone is the third important adrenal androgen and is formed from cholesterol through pregnenolone, from which 17-hydroxypregnenolone is formed, and then the side chain is removed. It is now known that this steroid is not exclusively adrenal in origin but can also be formed in small amounts both by the testis and the ovary. It can be converted to androst-4-ene-3,17-dione but the greater proportion becomes linked with sulphate so that dehydroepiandrosterone is present in the adrenal venous blood mainly as the sulphate conjugate. Other pathways for androgen synthesis exist such as through the formation of testosterone acetate, and these routes may possibly be important in virilizing disorders of the adrenal cortex.

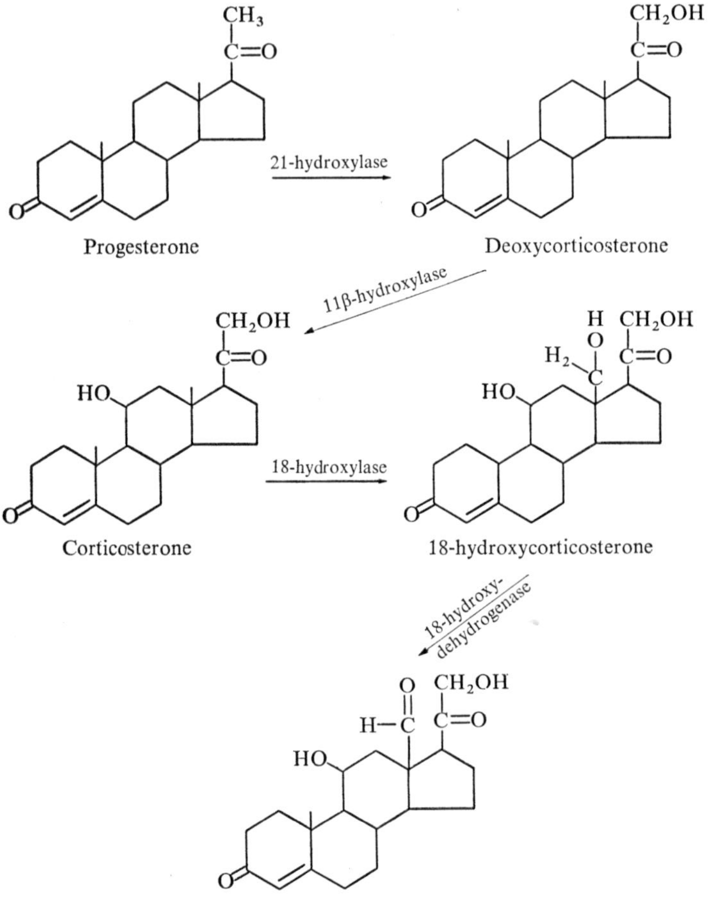

FIG. 3.2. The biosynthesis of aldosterone.

Biosynthesis of adrenal oestrogens (Fig. 3.4)

Evidence of oestrogen secretion by the human adrenal is not easily obtained, but in bovine adrenal preparations it takes place through the intermediate of androgen formation. One pathway involves the formation of androst-4-ene-3,17-dione, followed by the 19-hydroxy derivative, then 19-norandrostenedione and finally aromatization of ring A occurs to give oestrone.

Dorfman and Ungar (1965) survey the extensive literature relating to the biosynthesis of the adrenal steroids.

The action of ACTH

Steroidogenesis. ACTH increases the production of adrenocortical steroids within a few minutes of its administration although the synthesis of aldosterone is only temporarily stimulated. In spite of extensive investigation, there is still controversy as to the mechanism of its action. Stone & Hechter (1954) working with perfused bovine adrenals reported that ACTH had no effect

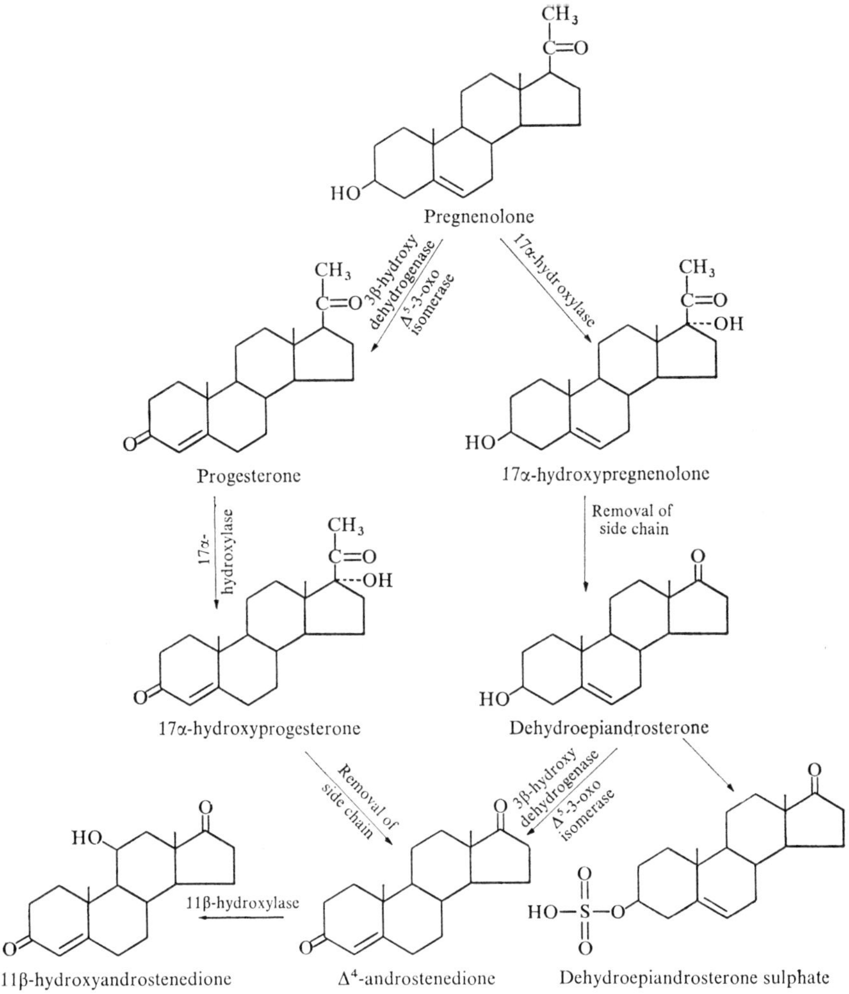

FIG. 3.3. The biosynthesis of adrenal androgens.

on steroid synthesis beyond the formation of progesterone and later work caused the site of action to be placed at the splitting of the side chain of cholesterol by the desmolase, thereby increasing the formation of pregnenolone, at this early rate-limiting stage of adrenal steroid synthesis.

Haynes & Berthet (1957) advanced another hypothesis of ACTH action which has received much attention. These workers found that ACTH specifically increased the amount of adrenal cyclic adenosine-3',5'-mono

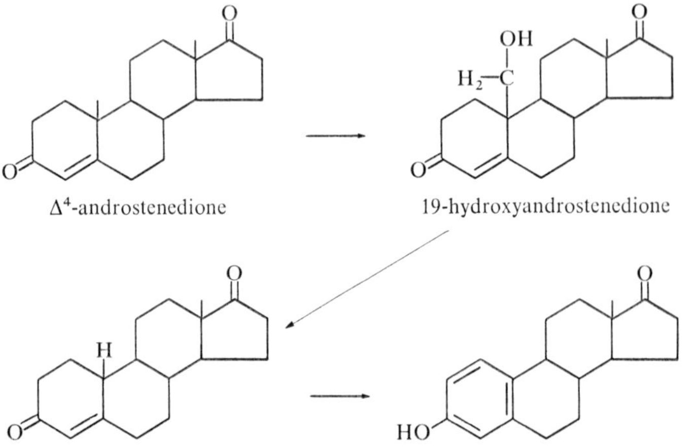

FIG. 3.4. Presumptive biosynthesis of oestrogen by human adrenal cortex.

phosphate (3',5'-AMP) which is an activator of phosphorylase. It was postulated that adrenal glycogen was converted to glucose-6-phosphate which was then metabolized by the hexose monophosphate shunt pathway to yield increased amounts of $NADPH_2$. This pyridine nucleotide is an essential cofactor for the formation of pregnenolone from cholesterol and also for the various hydroxylases concerned in the synthesis of cortisol and corticosterone. While there is general agreement that ACTH increases the concentration of 3'·5'-AMP, there are many objections to accepting the sequences postulated in the original hypothesis, such as a failure to reproduce the effect of ACTH by the addition of glucose-6-phosphate to adrenal preparations. It is unlikely that ACTH acts solely by increasing the amounts of $NADPH_2$. For example, cyclic 3',5'-AMP has been shown to increase the activity of 11β-hydroxylase in adrenal homogenates. There may be, therefore, a regulating effect of ACTH on several steps of steroid synthesis.

The addition of calcium ions or the freezing and thawing of adrenal tissue cause an increase in steroid synthesis *in vitro*. It is possible that these agents act by affecting the permeability of the mitochondria and there is evidence from electron microscope studies that ACTH has a similar effect.

Ascorbic acid depletion. The high concentration of ascorbic acid within the adrenal gland and the ability of ACTH to cause its rapid discharge have been known for many years. This depletion of ascorbic acid has been widely used as a biological assay of ACTH activity, and has led to speculation about its role in steroid synthesis. Careful time studies on the adrenal venous blood of the dog have shown that the release of ascorbic acid always precedes increased steroid secretion, the former appearing at about two minutes and the latter four minutes after ACTH. Jenkins (1962) could find no evidence of a direct effect of ascorbic acid on steroidogenesis from cholesterol *in vitro*, although there was some stimulation of 11β-hydroxylation only, under conditions of limited NADPH$_2$.

Nucleic acid and protein synthesis. In addition to the very rapid action of ACTH on steroid synthesis, which can be seen experimentally within a few minutes, the long-term effects of the hormone have also to be considered, and an increase in adrenal size is well known to occur after continued administration of ACTH or in Cushing's syndrome. Several workers have reported an increased nucleic acid synthesis, first RNA and then DNA. An increase in synthesis of adrenal protein under the influence of ACTH has been found less consistently.

The various actions of ACTH are fully discussed by Hilf (1965).

REFERENCES

DORFMAN, R. I. and UNGAR, F. (1965). *Metabolism of Steroid Hormones*, Academic Press, London.
HAYNES, R. C. and BERTHET, L. (1957). *J. Biol. Chem.* **225**, 115–124.
HILF, R. (1965). *New Eng. J. Med.* **273**, 789–811.
JENKINS, J. S. (1962). *Endocrinology.* **70**, 267–271.
POPJAK, G. and CORNFORTH, J. W. (1960). *Advan. Enzymol.* **22**, 281–335.
STONE, D. and HECHTER, O. (1954). Arch. Biochem. Biophys. **51**, 457–469.
WERBIN, H., CHAIKOFF, I. L., and JONES, E. E. (1960). *J. Biol. Chem.* **235**, 1629–1633.

4

THE METABOLISM OF THE ADRENAL STEROIDS

The removal of the adrenocortical hormones from the body is preceded by a series of changes in their chemical structure. These metabolites are formed mainly, but not exclusively, in the liver and the result is often a loss of the biological activity of the hormones. The following transformations in the adrenal steroids can occur.

Reduction of ring A. This alteration in the steroid structure accounts for the largest part of hormone metabolism. First, reduction of the 4,5 double bond takes place by an enzyme which is specific for each steroid. In man this enzyme is almost exclusively found in the liver and requires $NADPH_2$ as cofactor. The resultant C-5 hydrogen atom can be either α or β in its position. Two different enzymes are involved, and in studies on rat liver the 5α-reductase is found in the microsomes, whereas the enzyme responsible for 5β-reduction is present in the soluble fraction of the cell. The 5α- or 5β- dihydro-compound is then rapidly reduced at the 3-oxo group to form a 3-hydroxyl, usually α- in position, by an enzyme which is relatively non-specific, is not confined to the liver, and utilizes either $NADH_2$ or $NADPH_2$.

Reduction of the 20-oxo group. Reduction of the C-20 ketone by a 20-dehydrogenase enzyme occurs in many tissues and gives rise to a secondary alcohol of which α- or β-epimers exist. In the graphic formula the α-position of the hydroxyl group is on the right of C-20 and the β-position is on the left (Fig. 4.1).

Removal of the side chain. C_{21} compounds such as cortisol can lose their side chain oxidatively to form a C_{19} compound with a 17-oxo group.

Oxidation of the 11β-hydroxyl group. The 11β-hydroxyl is oxidized in various tissues to the 11-ketone. In the liver but not in other tissues of man this reaction is freely reversible.

6β-hydroxylation. Hydroxylation at the 6β-position may occur without any further change in the steroid structure and renders the compound water-soluble.

Conjugation. After reduction of ring A the steroid metabolites are conjugated

23

with either glucuronic acid or sulphate. The glucuronides or glucosiduronates, as they are more correctly, if rarely, called, are formed mainly in the liver by the transfer of glucuronic acid from uridine diphosphoglucuronic acid to the hydroxyl group of the steroid, usually at C-3, by means of the enzyme glucuronyl transferase present in liver microsomes. In the case of aldosterone, a

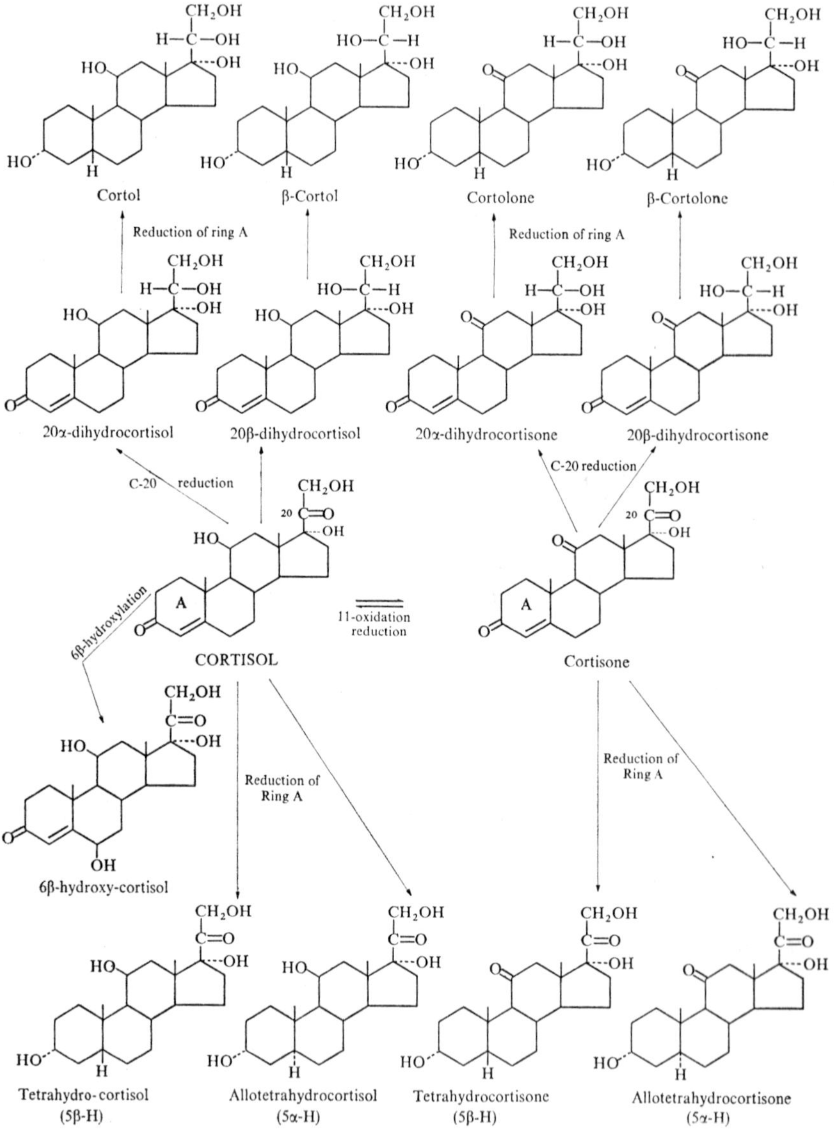

FIG. 4.1. C_{21} metabolites of cortisol.

glucosiduronate at C-18 is probably formed also. The sulphates occur mainly as conjugates of some C_{19} compounds, especially those with a 3β-hydroxyl group and here the sulphate donor is probably 3'-phosphoadenosine-5'-phosphosulphate. The presence of small amounts of cortisol and corticosterone-21-sulphates in human urine has also been reported. All these conjugates are very water-soluble and are therefore rapidly excreted by the kidney. In man the urine is the main route of excretion of the adrenal steroid metabolites, very little appearing in the bile and faeces.

The metabolism of cortisol (Figs. 4.1 and 4.2)

After the administration of a physiological amount of cortisol which is radioactively labelled, over 90% of the radioactivity is found in the urine within 48 hours, the majority being excreted within 24 hours. Very little, about 0·8%, is excreted as cortisol itself. The chief metabolites of cortisol are the tetrahydro-derivatives of cortisol and cortisone together with the cortols and cortolones (Fukushima et al. 1960).

Oxidation-reduction of the 11-oxy group. The fact that approximately two-thirds of the urinary metabolites of cortisol are present in the 11-oxo form

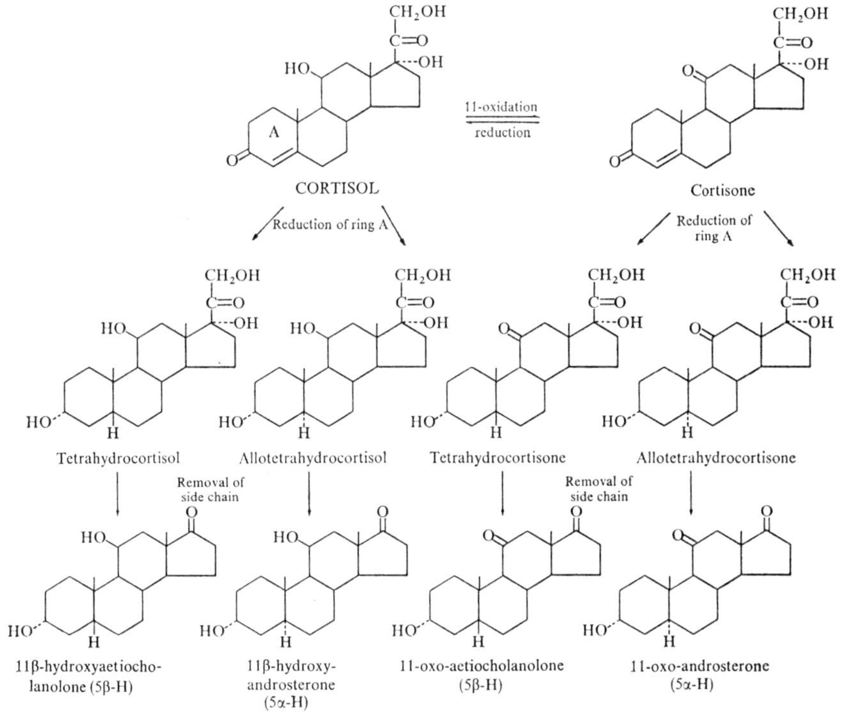

FIG. 4.2. C_{19} metabolites of cortisol.

indicates that there is considerable oxidation of the 11β-hydroxyl to form cortisone. This reaction takes place not only in the liver but also in other tissues, notably the kidney (Jenkins, 1966). The reverse reaction, whereby the 11-oxo group is reduced to the 11β-hydroxyl is very important since this is the mechanism by which 11-oxo steroids such as cortisone and prednisone are converted to the biologically active steroids cortisol and prednisolone. In man, the reduction of the 11-oxo group only takes place to a significant extent in the liver. Jenkins & Sampson (1967) found that the efficiency of this conversion is much greater for prednisone than for cortisone owing to the latter being more extensively metabolized to tetrahydro-compounds.

Reduction of ring A. First, the 4,5 double bond is reduced to dihydrocortisol which is then rapidly reduced further at the 3-oxo group to give a 3α-hydroxy, tetrahydro compound. The hydrogen at C-5 of tetrahydrocortisol is predominantly in the β-position, and only small amounts of the 5α-isomer, sometimes called allotetrahydrocortisol, are produced. Cortisone formed from cortisol is metabolized in a similar manner.

Reduction of the C-20 ketone. The formation of 20α- and 20β-dihydrocortisol takes place in many tissues but very little appears as such in the urine since these compounds rapidly undergo further reduction at ring A to form cortol and β-cortol. Cortolone and β-cortolone similarly arise from cortisone.

Tetrahydrocortisol, tetrahydrocortisone, the cortols and the cortolones are

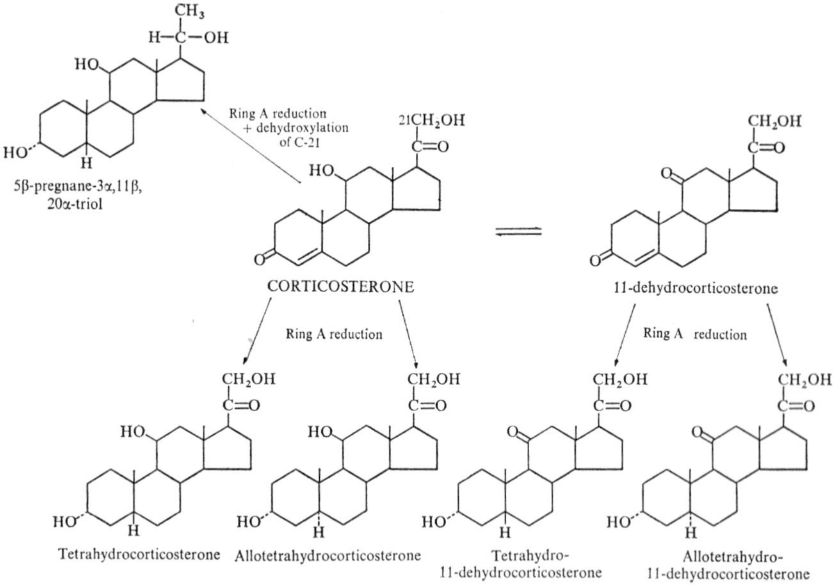

FIG. 4.3. Metabolites of corticosterone.

excreted in the urine as glucuronides, and together account for over 80% of the total glucuronide fraction.

Hydroxylation at the 6β-position. With the development of improved techniques for the extraction and separation of steroid metabolites it became apparent that there were present in urine very polar compounds which were not conjugated. The predominant unconjugated urinary metabolite of cortisol is 6β-hydroxycortisol and is formed mainly in the liver but also in the adrenal and kidney. It normally accounts for about 3% of the cortisol secreted but in circumstances where reduction of ring A does not occur to the full extent, notably in the new-born infant, 6β-hydroxycortisol assumes a much greater prominence as a cortisol metabolite. There is evidence that 2α-hydroxylation of cortisol can also take place, at least *in vitro*, since 2α-hydroxycrotisol has been siolated from human adrenal gland incubates. This substance is a major urinary metabolite in some strains of guinea pigs.

Metabolism of cortisol to 17-oxosteroids (Fig. 4.2). The conversion of cortisol to C_{19} compounds involves loss of the side chain and this appears to take place after reduction of ring A. Since the latter is predominantly 5β in orientation, the resultant 17-oxosteroid is 11β-hydroxy- or 11-oxoaetiocholanolone. Very little of the 5α-isomer, 11β-hydroxyandrosterone, is formed. From 2 to 12% of cortisol is converted to 17-oxosteroids and the adrenal origin of these 17-oxosteroids, as distinct from those from other sources is shown by the preservation of the 11-oxy substituent in the molecule.

Metabolism of corticosterone

The main reactions which take place in the metabolism of corticosterone are similar to those found in cortisol and are shown in Fig. 4.3. Oxidation of the 11β-hydroxyl occurs to form 11-dehydrocorticosterone and reduction of ring A occurs to give tetrahydro-derivatives of both compounds. In addition, reduction of the 21-hydroxyl has been reported to give the 21-deoxy-compound, 5β-pregnane-3α,11β,20α-triol.

Metabolism of aldosterone

When radioactive aldosterone is administered, over 90% of the radioactivity appears in the urine in 48 hours. About 60% of the radioactivity is present as metabolites conjugated with glucuronic acid. Quantitatively, reduction of ring A is the most important, to form tetrahydroaldosterone, in which the 18-aldehyde group remains intact. This metabolite comprises about 20–40% of the aldosterone secreted. The other major derivative of aldosterone in urine is of an unusual type, and consists of aldosterone conjugated with glucuronic acid, without prior reduction of ring A, and is sometimes known as the 3-oxo-conjugate of aldosterone. This, however, is a misnomer, since there is now evidence that the metabolite is a conjugate with glucuronic acid at the C-18 position of aldosterone (Underwood & Tait, 1964). Unlike the glucuronide conjugate of tetrahydroaldosterone, it is not specifically hydrolysed by the enzyme β-glucuronidase, but in fact readily hydrolyses, at room temperature, if brought to pH1, to yield free aldosterone. This acid-labile conjugate accounts for 5–15% of aldosterone secreted, and of this about half is formed extra-hepatically in the kidneys. This is the reason why more acid-labile conjugate is found in urine after the intravenous administration of aldosterone than when the oral route is used. When aldosterone is given orally it is metabolized completely by the liver and none appears in the peripheral blood to be presented to the kidney. The ready hydrolysis of this conjugate to give free aldosterone is the basis of many assay methods for the hormone. Only the naturally occurring dextroaldosterone is metabolized to tetrahydro- and acid-labile derivatives. Reduction of the side chain of aldosterone does not take place to any significant extent, and pairs of 11β-hydroxy,11-oxo compounds do not occur as in the case of cortisol.

Metabolism of adrenal androgens (Fig. 4.4)

Dehydroepiandrosterone

This C_{19} compound is known to be secreted into the adrenal venous blood mainly as the sulphate conjugate. After injection of the radioactive steroid about 1 to 12% is excreted unchanged into the urine, some of the sulphate is

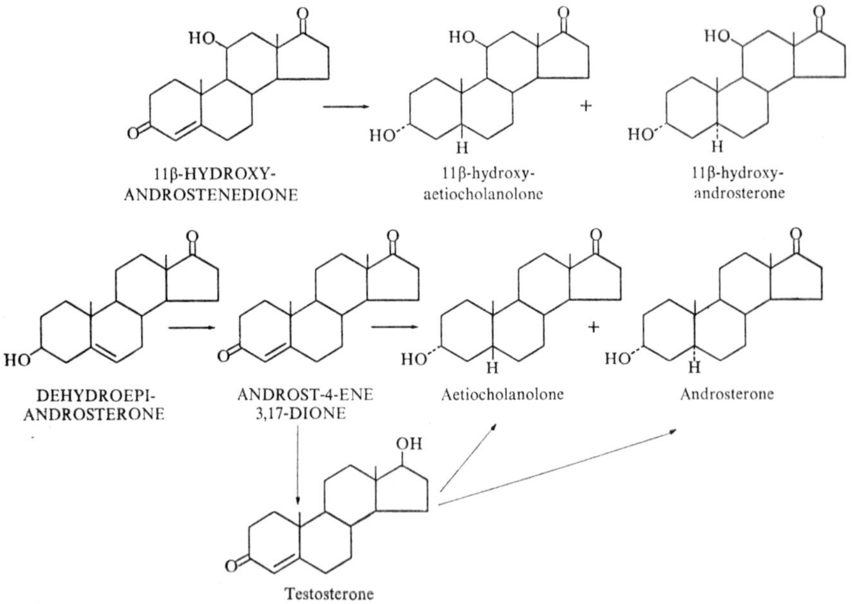

Fig. 4.4. Metabolites of adrenal androgens.

further metabolized in the conjugated form, but much is hydrolysed before undergoing several possible metabolic transformations. Oxidation of the 3β-hydroxyl group and isomerization of the C-5,6 to the C-4,5 double bond yields androst-4-ene-3,17-dione; thence it can form testosterone and the oestrogens. Alternatively androstenedione can be reduced at ring A to form a mixture of androsterone and aetiocholanolone.

Androst-4-ene-3,17-dione

This compound can be converted to testosterone, but the failure to find testosterone in normal adrenal venous blood in significant quantities means that this conversion usually takes place outside the gland. Reduction of ring A yields a mixture of androsterone and aetiocholanolone in approximately equal amounts.

11β-*hydroxyandrost-4-ene-3,17-dione*. Metabolism of this C_{19} compound involves reduction of ring A and some oxidation of the 11β-hydroxy group to the 11-ketone. The result is predominantly 11β-hydroxy- and 11-oxoandrosterone with lesser amounts of 11-oxy aetiocholanolone.

Summary of the factors which influence the orientation of ring A-reduction

Reduction of the 4,5 double bond of C_{21} compounds, e.g. cortisol, occurs predominantly in the 5β-position to give tetrahydrocortisol, and only to a minor extent in the 5α-position, allotetrahydrocortisol. Reduction of the 4,5 double bond of C_{19} compounds without a 11-oxy substituent, e.g. androstenedione, gives equal amounts of the 5α-metabolite androsterone, and the 5β-compound aetiocholanolone, but the presence of a 11-oxy-group in the C_{19} steroid orientates the predominant reduction to the 5α-position.

Metabolism of synthetic derivatives of cortisol

The effect of synthetic modifications of the cortisol molecule on its metabolism have been discussed in Chapter 2. In general, these modifications hinder those transformations which inactivate the biologically important groups. *Prednisolone and prednisone*. Owing to the angulation of the A/B ring junction caused by the presence of the 1,2 double bond, reduction in ring A is greatly diminished so that very little of the tetrahydro compounds is formed. Some prednisolone is converted to prednisone and they are both reduced at C-20 or excreted unchanged. The disappearance rate of prednisolone from the blood is about half that of cortisol.

9α-*Fluorocortisol*. The main effect on metabolism produced by the presence of the 9α-fluorine atom is the prevention of oxidation of the 11β-hydroxyl so that no 11-oxo derivatives are excreted, and also the tetrahydro derivatives contain a significantly higher proportion of 5α-H steroids.

The presence of a 2α-methyl group blocks oxidation and reduction at C-11 almost completely. The result is that whereas 2α-methyl cortisol is about 5 times as active as cortisol, 2α-methyl cortisone is virtually inactive, since it cannot be reduced to the 11β-hydroxy form.

Dexamethasone. In addition to the effects due to the 1,2 double bond and 9α-fluorine atom already described, this compound also has a 16α-methyl group which decreases reduction at the C-20 position.

Aspects of steroid metabolism concerning the synthetic analogues of cortisol are discussed by Bush (1962).

Rate of metabolism and the effect of extra-adrenal factors

The rate of metabolism of a steroid is best studied by following the disappearance of radioactivity from the plasma after the intravenous injection of a trace amount of labelled hormones. During the first 20 minutes there is a rapid drop in concentration owing to complex equilibrations taking place within extracellular and intracellular compartments (Tait *et al.* 1962). Thereafter, the disappearance of the steroid from the plasma falls logarithmically against time. The normal slopes for cortisol and aldosterone plotted semi-logarithmically are shown in Fig. 4.5. If the slope of the disappearance rate is measured, the time required for half the plasma steroid to disappear

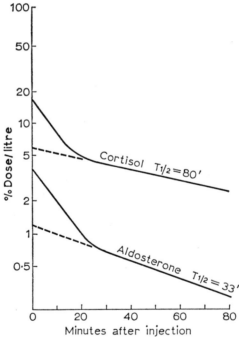

FIG. 4.5. Comparison of the plasma disappearance curves after a tracer dose of labelled cortisol and aldosterone (slightly modified from Tait *et al.* 1962)

can be determined. This biological half-life is a convenient measure of the rate of metabolism. The normal half-life of cortisol is 80 minutes, whereas that of aldosterone is only about 33 minutes, and, unlike cortisol, aldosterone is completely metabolized after a single passage through the normal liver. Tait *et al.* (1962) have shown that the turnover of aldosterone during 24 hours is approximately 10 times that of cortisol. These differences are largely due to the much greater degree of protein binding of cortisol compared with that of aldosterone. The metabolism of cortisol is therefore affected by changes in protein binding so that when this is increased, as in pregnancy or after the administration of oestrogens, the half-life of cortisol is greatly increased. The effect of pregnancy on aldosterone metabolism is of a different kind. It has been reported that 50% of the total metabolites is represented by the acid-labile conjugate during the latter part of pregnancy instead of the normal 5 to 15%. Metabolism of the adrenocortical hormones is also affected by the thyroid gland (see Chapter 8).

The effects of adrenal steroid metabolism

The metabolism and conjugation of the adrenal steroids facilitate their excretion. Since the level of steroid in the blood is normally maintained within a certain range, the liver and other tissues which metabolize the hormones can be regarded as regulating their secretion. In many cases the chemical transformation of the steroid leads to a loss of biological activity, either partial or more often complete. There are, however, some metabolites which in fact have different physiological actions from those of the parent steroids. Examples are aetiocholanolone and androsterone, which are metabolites of testosterone, androst-4-ene-3,17-dione, and dehydroepiandrosterone. Aetiocholanolone has a marked pyrogenic effect in man when injected intramuscularly and an excess level of free aetiocholanolone in the blood has been held to be a cause of periodic fever (Bondy *et al.* 1965). The 5α-isomer, androsterone, has no pyrogenic action but its administration is followed by a fall in the level of plasma cholesterol. The effect of thyroid hormone both on reducing plasma cholesterol and also on increasing the formation of androsterone, is therefore of interest.

Although the suggestion has been frequently made that the metabolism of steroids is directly concerned with their actual mode of action there is no real evidence for this in the case of the adrenal hormones. On the other hand, many of the chemical transformations which lead to inactivation of the adrenal hormones take place in peripheral tissues, and it seems possible that metabolism of steroids by these tissues may determine the concentration of biologically active hormone which reaches the cells.

A very comprehensive account of steroid metabolism is to be found in the monograph by Dorfman & Ungar (1965).

32 Biochemical aspects of adrenal cortex

REFERENCES

BONDY, P. K., COHN, G. L., and GREGORY, P. B. (1965). *Medicine*, **44**, 249–262.
BUSH, I. E. (1962). In *The Human Adrenal Cortex*. pp. 138–171. Ed. by A. R. CURRIE, T. SYMINGTON, and J. K. GRANT. Livingstone, Edinburgh.
DORFMAN, R. I. and UNGAR, F. (1965). *Metabolism of Steroid Hormones*. Academic Press, London and New York.
FUKUSHIMA, D. K., BRADLOW, H. L., HELLMAN, L., ZUMOFF, B., and GALLAGHER, T. F. (1960). *J. Biol. Chem.* **234**, 2246–2252.
JENKINS, J. S. (1966). *J. Endocr.* **34**, 51–56.
JENKINS, J. S. and SAMPSON, P. A. (1967) *Brit. Med. J.* **2**, 205–207.
TAIT, J. F., TAIT, S. A. S., LITTLE, B., and LAUMAS, K. (1962). In *The Human Adrenal Cortex*. pp. 107–123. Ed. by A. R. CURRIE, T. SYMINGTON, and J. K. GRANT. Livingstone, Edinburgh.
UNDERWOOD, R. H. and TAIT, J. F. (1964). *J. clin. Endocr.* **24**, 1110–1124.

5

THE CONTROL OF ADRENOCORTICAL
SECRETION

If the pituitary gland is removed from an animal, the level of cortisol in the blood rapidly falls to very low levels. The polypeptide hormone, adreno-corticotrophin (ACTH), secreted by the anterior lobe of the pituitary, is responsible for maintaining and increasing cortisol secretion when required. The normal concentration of ACTH in the plasma is not easy to determine but when very sensitive bioassays have been used, values of approximately 0·25 milli-units per 100 ml., have been reported.

The chemical nature of ACTH

ACTH has been obtained in purified form from the pituitaries of many species, the amino acid sequences have been determined, and now total synthesis has been achieved. The structure of the natural hormone is that of a straight-chain polypeptide consisting of 39 amino acids (Fig. 5.1) of which the first 24 are similar for the pig, sheep, beef, and human hormones. (Li, 1962). Species differences are to be found in positions 25–33. Since the presence of amino acids 1–24 only is adequate for full adrenocorticotrophic action, there is little difference in activity between the various animal and human hormones. Synthetic ACTH consisting of the necessary 1–24 amino acids is now commercially available, and has the therapeutic advantage over the ACTH prepared from animal pituitaries of a reduced risk of producing allergic reactions.

Melanocyte-stimulating hormone (MSH)

This hormone, the action of which is to increase melanindispersion within the melanocytes of the skin, is closely associated with ACTH and early work rather confused the identity of the two. MSH has been isolated in two forms. α-MSH has a structure which is closely similar to the first 13 amino acid sequences of ACTH and is common to all species. β-MSH comprises a group of polypeptides which are species-specific. Amino acids 4–10 are common to ACTH and to both α-and β-MSH, so that some overlap of activities occurs.

```
 1   2   3   4   5   6   7   8   9  10  11  12  13  14  15  16  17  18  19  20  21  22  23  24
Ser-Tyr-Ser-Met-Glu-His-Phe-Arg-Tyr-Gly-Lys-Pro-Val-Gly-Lys-Lys-Arg-Arg-Pro-Val-Lys-Val-Tyr-Pro
```

```
        25  26  27  28  29  30  31  32  33
Beef    Asp-Gly-Glu-Ala-Glu-Asp-Ser-Ala-Gluta
Sheep   Ala-Gly-Asp-Asp-Glu-Ala-Ser-Gluta
Pig     Asp-Gly-Ala-Glu-Asp-Glu-Leu-Ala-Glu
Man     Asp-Ala-Gly-Glu-Asp-Gluta-Ser-Ala-Glu
```

```
 34  35  36  37  38  39
Ala-Phe-Pro-Leu-Glu-Phe
```

Ala = Alanine
Arg = Arginine
Asp = Aspartic acid
Glu = Glutamic acid
Gluta = Glutamine
Gly = Glycine
His = Histidine
Leu = Leucine
Lys = Lysine
Met = Methionine
Phe = Phenylalanine
Pro = Proline
Ser = Serine
Tyr = Tyrosine
Val = Valine

FIG. 5.1. The structure of ACTH in different species

ACTH secretion within the pituitary has generally been considered to be localized to the basophil cells, but the identification of the various cell types in the human pituitary is not easy and a complex nomenclature to include many sub-groups has been devised. The secretion of large amounts of ACTH by certain pituitary tumours whose structure would ordinarily be classified as chromophobe adds to the confusion. The subject is reviewed by Purves (1966).

THE CONTROL OF ACTH SECRETION BY THE CENTRAL NERVOUS SYSTEM

For many years it had been known that psychological disturbances such as anxiety, or physical trauma of various kinds, collectively termed "stress" produced a rapid increase in adrenocortical secretion. In animal experiments, destruction of the ventral hypothalamus, particularly that part known as the median eminence, removes the pituitary-adrenal response to stress, and the lesion is followed by a fall in plasma cortisol. In both man and animals, section of the pituitary stalk has similar effects, but if revascularization of the pituitary takes place, recovery of pituitary-adrenal function occurs. The essential anatomical relationship of the hypothalamus to the pituitary is well shown by animal experiments in which the pituitary is removed and grafted into another site such as the capsule of the kidney. Even if revascularization is successful, normal pituitary control of the adrenal does not occur, but it can be reimposed if the pituitary is again transplanted back to its proper site under the median eminence region of the hypothalamus.

THE MEDIAN EMINENCE AND ACTH-RELEASING HORMONE

The median eminence occupies the central area of the base of the hypothalamus and lies in the floor of the third ventricle. It is composed of specialized neural tissue and through it pass the nerve tracts from the supra-optic and paraventricular nuclei to the posterior lobe of the pituitary. No neural connections can be detected between the hypothalamus and the anterior lobe, but the anatomy of the blood supply to this part of the pituitary is unusual. A primary plexus of capillaries ramifies round nerve endings in the median eminence, and from thence the vessels pass down to end in the anterior lobe of the pituitary, thus constituting a portal system (Fig. 5.2). The present concept of the control of anterior pituitary function is based upon the presence of its peculiar blood supply. There is now good evidence that various neurohormones are secreted by the nerve endings in the median eminence, and pass down the portal vessels to the anterior pituitary to cause a release of the approriate trophic hormone. The chemical nature of ACTH (corticotrophin)-releasing factor (CRF) has not been completely elucidated but may be a polypeptide somewhat similar to vasopressin. For a long time there was considerable controversy about the relationship of vasopressin to CRF and some workers believed the two neurohormones to be identical. It is now certain that while vasopressin, in very large doses, can lead to the release of ACTH, it is not the physiological CRF. The origin of the nerve fibres in the

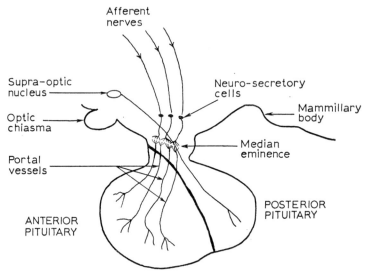

Fig. 5.2. Diagramatic representation of the hypothalamic control of the pituitary

median eminence responsible for the secretion of CRF is not known, but they probably connect with the supra-optic and paraventricular nuclei in the hypothalamus and also with more distant areas. In fact, through its vascular connections, the median eminence appears to provide the final link between the central nervous system as a whole and the anterior lobe of the pituitary. That this view may be an oversimplification is suggested by the work of Egdahl (1961) who found in dogs that if all brain tissue was removed down to the pons leaving an isolated pituitary–hind brain preparation, ACTH release was increased. It was suggested that normal inhibitory influences, perhaps from the cerebral cortex, had been removed in this preparation.

THE "FEEDBACK" CONTROL OF CORTISOL SECRETION

One of the important side effects which follow the administration of of glucocorticoids in a dosage higher than the physiological requirement, is suppression of ACTH release and subsequently adrenal atrophy. Conversely, when cortisol secretion is deficient after adrenalectomy or in Addison's disease, the level of ACTH in the blood rises to easily measurable levels and falls again within two hours after the administration of small amounts of cortisol. Furthermore, plasma levels of cortisol are maintained within the normal range under conditions of either increased or decreased rates of metabolism, owing to an accompanying increase or decrease in secretion rate. Thus, the actual level of cortisol in the blood determines the release of ACTH by the pituitary. Earlier studies suggested that cortisol acted directly on the pituitary itself, but it now appears that the site of this "feedback" control lies in the hypothalamus. The implantation of a potent corticosteroid such as dexamethasone into the median eminence leads to a fall in adrenal secretion of cortisol. Lesions involving the hypothalamus are followed by a failure of the pituitary to respond to decreased cortisol levels induced by the drug metyrapone, and by a failure of exogenous corticosteroids to cause a fall in cortisol secretion.

The relationship between the central nervous system and feedback control of cortisol secretion

The rise in cortisol which rapidly takes place in response to stress would seem to be acting antagonistically to the feedback mechanism, which should then operate to reduce ACTH release. Yates & Urquhart (1962) have attempted to integrate the two mechanisms by postulating that under conditions of stress the "set-point" of the feedback control is simply reset at a higher level. Against this hypothesis is the difficulty of suppressing the increase in cortisol levels following severe stress even with pharmacological doses of glucocorticoids. Moreover, there are many instances of patients with hypothalamic-pituitary lesions in whom tests of the response to stress and the feedback control give divergent results. The bulk of the evidence is against there being a single controlling mechanism for the regulation of pituitary-adrenal function. The subject is well reviewed by Reichlin (1963).

The diurnal variation in cortisol secretion

With the development of sensitive methods for the estimation of corticosteroids in the blood it became apparent that the level of cortisol was not constant throughout 24 hours but showed a diurnal variation or circadian rhythm, as it is sometimes called (Fig. 5.3). Blood cortisol is normally at the maximum concentration at about 8.00 a.m., and falls during the course of the day to reach its lowest point at midnight or soon after, following which it begins to rise sharply at about 4.00 a.m. This diurnal variation is dependent upon the integrity of the central nervous system since it may be abolished in the presence of some brain lesions, especially those centred around the hypothalamus. Light does not seem to be a factor since blind persons preserve the rhythm. Reversal of the sleep period, however, if continued for more than a few days, as in night workers, results in a reversed diurnal rhythm. It seemed likely that the diurnal variation was due to a periodicity in the release of ACTH, and by the use of a very sensitive assay for ACTH Ney *et al.* (1963) have demonstrated directly that there is a diurnal variation in the release of ACTH, the mean normal plasma values being 0·25 milli-units per 100 ml. at 6.00 a.m. and 0·11 milli-units per 100 ml. at 6.00 p.m. The variation

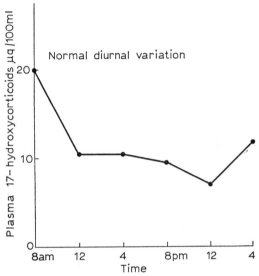

FIG. 5.3. The normal diurnal variation in plasma cortisol (17-hydroxycorticosteroids)

in cortisol levels throughout the 24 hours must be considered when comparing levels in different samples of plasma so that the time of obtaining the speciments should be recorded. Loss of a diurnal rhythm may be of diagnostic importance in hypothalamic lesions and in Cushing's syndrome.

The control of adrenal androgen secretion

Although ACTH will stimulate the output of adrenal C_{19} compounds as measured by urinary 17-oxosteroid excretion, the increase is often variable when compared with that of cortisol. It is relatively small in children and is sometimes greater than normal in women with hirsutism. These findings could be due to differences in the response of the adrenal cortex, but the reduced response of 17-oxosteroid excretion to ACTH in patients with hypopituitarism whose cortisol response was adequate, led Mills *et al.* (1962) to suggest that there is a further pituitary hormone which acts synergistically with ACTH in the control of adrenal androgen secretion. There is as yet no definite proof of the existence of such a hormone.

THE CONTROL OF ALDOSTERONE SECRETION

Early studies appeared to indicate that the pituitary played little or no part in the control of aldosterone secretion. In fact, the administration of ACTH will produce an increase in the output of aldosterone, although the amount required is about 10 times that which causes maximal cortisol secretion, and the effect is transitory, even if the administration of ACTH is prolonged. It seems probable that ACTH does have a role in the regulation of aldosterone secretion, particularly in conditions of severe stress, but its relative importance is not yet settled. It is certain, however, that even in the absence of the pituitary, considerable stimulation of aldosterone output can be achieved. The most important factors responsible for an increase in secretion are depletion of sodium; haemorrhage and conditions which result in a reduced plasma volume; constriction of the thoracic inferior vena cava; loading with potassium, especially in potassium-depleted subjects. The first three of these agents can be shown to be dependent upon the presence of the kidneys for their action, although if the fall in plasma concentration of sodium is considerable it may be effective even in the bilaterally nephrectomized animal. Loading with potassium also acts independently of the kidney. In the dog, a rise in plasma potassium of 1·3 m-equiv. per litre, or a fall in sodium as great as 14 m-equiv. per litre have a direct stimulatory effect on the adrenal cortex. The renal factor which is responsible for the stimulatory effect of a low sodium diet or a reduced plasma volume has been identified as renin, acting through the formation of angiotensin in the plasma.

The renin-angiotensin mechanism

The infusion of renin or angiotensin into man and experimental animals, in doses which have little pressor effect, is followed by a sustained increase in aldosterone secretion. In man no accompanying increase in the cortisol level is seen. The action of angiotensin is directly on the adrenal since it is fully effective when injected into the arterial blood supply of the gland. After haemorrhage and constriction of the inferior vena cava, an increase in the renin concentration of the renal vein blood has been demonstrated. After sodium depletion there is an increased granularity of the juxta-glomerular cells of the kidney which are probably the source of renin, and increased amounts of renin are found in the peripheral blood. Posture is also known to affect aldosterone secretion. It has been reported that on assuming the upright position there is a fall in renal arterial perfusion pressure which leads to a rapid release of renin and this is followed by increased aldosterone production (Gordon et al. 1966). The elevation of renin which is known to occur in malignant hypertension may explain the high aldosterone secretion which is found in this particular type of the disease.

D

The effect of the central nervous system

Farrell (1964) as a result of a series of experiments carried out on the dog, claimed that the central nervous system controlled aldosterone secretion, not through the median eminence, but through a hormone elaborated by the pineal and the adjoining posterior diencephalon. It was later reported that the pineal also produced an inhibitory hormone. These findings have not been universally accepted, and it is not possible at present to state the physiological role, if any, of the central nervous system in the regulation of aldosterone secretion. In man it seems likely that changes in sodium concentration and in blood volume acting through the renin-angiotensin mechanism are mainly responsible for the control of aldosterone, with ACTH acting in a subsidiary capacity, but the evidence is still incomplete and it is quite possible that other regulatory processes will be discovered. The control of aldosterone secretion is discussed by Ganong *et al.* (1966).

REFERENCES

EGDAHL, R. (1961). *Endocrinology*, **68,** 574–586.
FARRELL, G. (1964). In *Aldosterone.* pp. 243–249. Ed. by E. E. BAULIEU and P. ROBEL. Blackwell, Oxford.
GANONG, W. F., BIGLIERI, E. G., and MULROW, P. J. (1966). *Recent Progr. Hormone Res.* **22,** 381–430.
GORDON, R. D., WOLFE, L. K., ISLAND, D. P., and LIDDLE, G. W. (1966). *J. clin. Invest.* **45,** 1587–1592.
LI, C. H. (1962). *Recent Progr. Hormone Res.* **18,** 1–40.
MILLS, I. H., BROOKS, R. V., and PRUNTY, F. T. G. (1962). In *The Human Adrenal Cortex.* pp. 204–216. Ed. by A. R. CURRIE, T. SYMINGTON, and J. K. GRANT. Livingstone, Edinburgh.
NEY, R. L., SHIMIZU, N., NICHOLSON, W. E., ISLAND, D. P., and LIDDLE, G. W. (1963). *J. clin. Invest.* **42,** 1669–1677.
PURVES, H. D. (1966). In *The Pituitary Gland*, vol. 1. pp. 147–232. Ed. by G. W. HARRIS and B. T. DONOVAN. Butterworths, London.
REICHLIN, S. (1963). *New. Eng. J. Med.* **269,** 1182–1191, 1246–1250, 1296–1303.
YATES, F. E. and URQUART, J. (1962). *Physiol. Rev.* **42,** 359–443.

6

THE BIOLOGICAL ACTIONS OF THE ADRENOCORTICAL HORMONES

Unlike the adrenal medulla, the adrenal cortex is essential for life, and complete adrenal cortical failure due to disease or surgical removal, is rapidly fatal in the absence of effective treatment. In man cortisol is especially concerned with this life-preserving function, and a closely associated but ill-understood phenomenon is the necessity for the presence of cortisol in adequate amounts to enable a patient to withstand any form of stress. If the supply of cortisol is insufficient, an otherwise non-fatal illness or trauma may lead to death.

The true physiological role of the adrenal hormones is seen when changes which accompany adrenal insufficiency are restored to the normal state by the administration of physiological amounts of hormone. These effects are to be distinguished from those observed after very large doses of corticosteroids have been given in the treatment of various non-endocrine diseases. Under these conditions either exaggerated responses occur, or additional, untoward effects may appear.

Cortisol and steroids with similar activity are known as "glucocorticoids", a somewhat misleading term since it includes not only the effect on glucose metabolism, but also the many other actions characteristic of this type of steroid.

Aldosterone is the natural example of what is called a "mineralocorticoid" since its main action is directed at the regulation of sodium and potassium excretion by the kidney. Deoxycorticosterone and fluorocortisol are other examples of mineralocorticoids. The former is not a normal secretory product in man, and the latter is an entirely synthetic compound, which also has some glucocorticoid activity.

Electrolyte metabolism and renal function

In severe adrenal insufficiency there is an excessive urinary excretion of sodium and a decreased excretion of potassium with a fall in glomerular filtration rate and an inability to produce a diuresis after an increased intake

41

of water. The administration of aldosterone or other mineralocorticoid restores to normal the plasma sodium and potassium concentrations by facilitating the renal reabsorption of sodium and the excretion of potassium and hydrogen ion in the distal convoluted tubule of the kidney. There is evidence that aldosterone also promotes the renal excretion of magnesium in the adrenalectomized patient (Horton & Biglieri, 1962). In addition, aldosterone has effects on cation transport outside the kidney, so that it decreases the concentration of sodium and increases that of potassium in the saliva and in the faeces. The administration of deoxycorticosterone, which has only mineralocorticoid activity, does not fully restore the reduced glomerular filtration rate in Addison's disease, and cannot correct the failure to excrete a water load. Cortisol also has considerable effects on renal function, and, in particular, will restore the ability to excrete water. This effect may be partly due to a further improvement in renal blood flow and glomerular filtration rate, but in addition there is probably a direct action of cortisol on the diluting segment of the nephron so that the permeability of the renal tubule to water is decreased. In this respect the action is opposite to that of vasopressin. In large, unphysiological doses of 200 mg or more daily, cortisol causes some retention of sodium and excretion of potassium.

The effects of cortisol and aldosterone on the renal tubule are discussed by Yunis *et al.* (1964).

Carbohydrate metabolism

In Addison's disease, there is a tendency to fasting hypoglycaemia, and the patient is unduly sensitive to the effects of insulin. Conversely, Cushing's syndrome is often associated with a fasting hyperglycaemia or an impaired glucose tolerance. Early work by Long *et al.* (1940) showed that, even in the fasting state, rats could deposit glycogen in the liver under the influence of cortisol. The extra glycogen was derived from newly synthesized glucose, and Long observed that this was accompanied by a rise in nitrogen excretion due to increased catabolism of protein. The stimulating effect of cortisol on gluconeogenesis is now well recognized and considerable work has been done to determine the mechanism of action.

It seems probable that several sites are involved; an inhibition of pyruvate oxidation to CO_2, an increased fixation of CO_2 by pyruvate, an increased trapping of amino acids by the liver, and an actual increase in the synthesis of hepatic enzymes such as glucose 6-phosphatase and fructose-1,6-diphosphatase.

In addition to stimulating gluconeogenesis, some workers claim that cortisol inhibits the utilization of glucose by the peripheral tissues. While this can be demonstrated in animals by the use of large amounts of hormone given for several hours, Jenkins *et al.* (1964) could find no inhibition of glucose uptake by the tissues of the human forearm over a period of three hours, when cortisol was infused intra-arterially in physiological amounts.

The effect of adrenal steroids on carbohydrate metabolism is reviewed by Landau (1965).

Protein metabolism

It has been shown that the effect of cortisol on protein metabolism is generally catabolic in most tissues although in the liver protein synthesis has been shown to take place. In Cushing's syndrome considerable wasting of muscle occurs together with loss of protein from the skin and bones leading to thinning of the skin and osteoporosis. Experimentally, in the rat cortisol increases the level of amino acids in the plasma and facilitates their uptake by the liver, whereas adrenalectomy increases the incorporation of amino acids into muscles. Cortisol increases the amino peptidase activity of muscle and also hepatic glutamic-pyruvic transaminase. These effects could be concerned with the regulation of gluconeogenesis, but a rise in hepatic glucose has been observed before there is evidence of protein catabolism so that additional mechanisms are involved in the initial synthesis of glucose.

These and other aspects of the metabolic effects of adrenal hormones are the subject of Ciba Foundation Study Group No. 6 (1960).

Fat metabolism

Knowledge of the effects of cortisol on fat metabolism is especially confused, mainly owing to the difficulty in separating the primary action of the hormone from effects which are secondary to those on carbohydrate metabolism. Thus, Cushing's syndrome is usually associated with increased fat deposits, but this is almost certainly due to a secondary increase in insulin secretion brought about by gluconeogenesis. Jenkins *et al.* (1964) in experiments on the human forearm showed that the local intra-arterial infusion of cortisol was followed by a release of free fatty acids from adipose tissue and this could be reversed by the simultaneous infusion of insulin.

The relationship of the adrenal cortical hormones to ketosis is also controversial but it is known that the administration of even small doses of cortisone to patients with both Addison's disease and diabetes mellitus may lead to severe ketosis. Scow & Chernick (1960) showed that glucocorticoids cause the rapid development of severe ketosis in the adrenalectomized pancreatectomized rat. In patients with intact pancreatic function, there is no effect on fasting ketosis when glucocorticoids are administered acutely, whereas in diabetic patients there is a rapid increase in ketonaemia together with an increase in plasma free fatty acids.

Blood pressure

Hypotension is an important clinical feature of Addison's disease but the relationship of adrenal cortical function to the maintenance of normal blood pressure is complex. The mineralocorticoids aldosterone, deoxycorticosterone, and fluorocortisol can restore the blood pressure to normal in Addison's disease. There is an increase in the total blood volume and the retention of sodium ion which is produced may also have a direct effect on the vascular musculature. Although glucocorticoids may also play some part, it has been shown in experimental animals that the mineralocorticoid deoxycorticosterone is much more effective than dexamethasone in potentiating the vascular response to noradrenaline (Schmid et al. 1967).

Blood cells

Adrenal insufficiency is usually associated with some degree of eosinophilia which falls rapidly after the administration of cortisol. The fall in the eosinophil count four hours after stimulation of cortisol secretion by ACTH was originally used by Thorn as a test of a normal adrenal response. Cortisol also causes a fall in lymphocytes and a reduction in lymphoid tissue, especially the thymus (Makman et al. 1967). These lytic effects on lymphoid tissue contrast strongly with the synthesis of new protein which is induced in liver cells. Neutrophil leucocytes tend to increase under the influence of cortisol.

Central nervous system

The presence of mental changes such as apathy or restlessness in Addison's disease is well recognized and they are associated with abnormalities in the electroencephalogram. These changes are reversed by the administration of a glucocorticoid but not a mineralocorticoid. Cushing's syndrome or the administration of large doses of cortisol is also sometimes complicated by mental disturbances, which may amount to psychosis, and usually improve after appropriate treatment of the hyper-adrenal state. Henkin & Bartter (1966) have observed that patients with Addison's disease have a greatly heightened sense of smell and taste which returns to normal only with the administration of glucocorticoids.

Anti-inflammatory action of glucocorticoids

The first reported success of cortisol in the treatment of rheumatoid arthritis showed that one of the most important aspects of glucocorticoid action from the therapeutic point of view is the anti-inflammatory effect. Vascular permeability is decreased, and the migration of inflammatory cells is reduced. As a result, there is an increased susceptibility to infection, and the spread of micro-organisms is facilitated. Depression of antibody formation has also been reported in some animals such as the rat and rabbit, but not in man. More recently it has been reported by Weissman & Thomas (1964) that glucocorticoids prevent the disruption of lysosomes. These are intra-cellular particles containing proteolytic enzymes which are known to be liberated by many agents capable of causing damage to tissues. It should be stated that all these effects on inflammatory reactions are seen after the use of large doses of cortisol, so that it is doubtful to what extent they represent a physiological action of the hormone.

The action of adrenal cortical hormones on the cell

The biological actions of the adrenal steroids described above, refer to the gross metabolic events which occur when glucocorticoids or mineralocorticoids are administered to the whole patient or animal. The fundamental action of the hormones must, however, lie on or within the cell. There has been considerable discussion as to whether a single common action such as alteration in cell permeability can explain the many diverse effects of the hormones, but any hypothesis must also account for the specificity of hormone action at different tissue sites. It has been suggested that the steroid hormones act as cofactors to enzymes or activate intra-cellular enzymes. There is little evidence for this in the case of the corticosteroids. One of the most fascinating recent developments in the field of hormone action is the suggestion that hormones may function as regulators of genes in the cells of the target tissues. The rapid growth of knowledge in molecular biology has provided the stimulus for this approach. It is claimed that cortisol induces enzyme formation in the liver by stimulating new messenger RNA synthesis. In the investigation of the action of aldosterone, Edelman has produced evidence that this hormone is preferentially localized in the cell nuclei. He maintains that aldosterone increases sodium transport, at least in the toad bladder, by stimulating DNA-dependent synthesis of RNA which leads to the production of enzymes involved in sodium transport. Experiments in this field are often open to several interpretations, and caution is necessary before accepting as fact the direct action of corticosteroids as genetic regulators, but the further exploration of steroid and gene activity promises to be of extreme interest. The subject is well reviewed by Hechter & Halkerston (1965) and the mode of action of aldosterone is discussed by Sharp & Leaf (1966).

REFERENCES

CIBA FOUNDATION STUDY GROUP No. 6 (1960). *Metabolic effects of Adrenal Hormones.* Churchill, London.
HECHTER, O. and HALKERSTON, I. D. K. (1965). *Ann. Rev. Physiol.* **27**, 133–162.
HENKIN, R. I. and BARTTER, F. C. (1966). *J. clin. Invest.* **45**, 1631–1639.
JENKINS, J. S., LOWE, R. D., and TITTERINGTON, E. (1964). *Clin. Sci.* **26**, 421–427.
LANDAU, B. R. (1965). *Vitam. and Horm.* **23**, 1–59.
LONG, C. N. H., KATZIN, B., and FRY, E. G. (1940). *Endocrinology*, **26**, 309–344.
MAKMAN, M. H., NAKAGAWA, S., and WHITE, A. (1967). *Recent Progr. Hormone Res.* **23**, 195–227.
SCHMID, P. G., ECKSTEIN, J. W., and ABBOUD, F. M. (1967). *J. clin. Invest.* **46**, 590–598.

Scow, R. O. and Chernick, S. S. (1960). *Recent. Progr. Hormone Res.* **16**, 497–545.
Sharp, G. W. G. and Leaf, A. (1966). *Physiol. Rev.* **46**, 593–633.
Weissman, G. and Thomas, L. (1964). *Recent Progr. Hormone Res.* **20**, 215–245.
Yunis, S. L., Bercovitch, D. D., Stein, R. M., Levitt, M. F., and Goldstein, M. H. (1964). *J. clin. Invest.* **43**, 1668–1676.

PART II

Disorders of Adrenocortical Function

7

DISORDERS OF BIOSYNTHESIS OF THE ADRENAL STEROIDS

A failure in the normal production of adrenal steroids occurs when there is a loss of secretory cells but biosynthesis is also impaired when there is a defect in one or more of the essential enzyme systems.

ADDISON'S DISEASE

In Addison's disease, whether it be due to auto-immune disease, tuberculosis, or any of the less common aetiological factors, there is a destruction of adrenocortical tissue so that the biosynthesis of all adrenal steroids is greatly reduced, and sometimes falls to negligible values. In many instances cortisol levels in the blood are close to zero, and 17-hydroxycorticosteroids in the urine are low. Urinary 17-oxosteroids are very low in females and in males they are reduced to about one third of normal. It is now known, however, that an appreciable number of cases of Addison's disease have a partial adrenal lesion only, and these patients may have blood and urinary corticosteroids which are within the normal range. The condition is recognized by the failure of the steroid levels to increase normally after the administration of ACTH, since the adrenal remnant is already under maximal stimulation by endogenous ACTH. An insufficient synthesis of cortisol leads to excessive release of ACTH through the feedback control, so that the levels of ACTH in the blood may reach 55 milli-units per 100 ml. compared with the normal morning value of about 0·6 milli-units obtained by Vance et al. (1962). The marked hyperpigmentation usually seen in these patients is an indication of the increased amount of MSH which is also present in the blood. The occurrence of fasting hypoglycaemia and the failure to excrete water in a normal manner are also manifestations of cortisol insufficiency and are discussed in more detail in Chapter 6. Aldosterone secretion, in common with that of cortisol, is also greatly reduced in Addison's disease, and this is followed by the renal loss of sodium, retention of potassium, and a reduction in extracellular volume. In adrenal insufficiency secondary to hypopituitarism, these electrolyte changes are not usually apparent since in most cases the secretion of aldosterone is preserved.

The condition of selective hypoaldosteronism with normal cortisol production has been reported on rare occasions (Posner & Jacobs, 1964).

CONGENITAL ADRENAL HYPERPLASIA

A variety of congenital enzymatic defects in the synthesis of cortisol have now been recognized. The deficiency of cortisol leads to an increased release of ACTH with the result that adrenals become hyperplastic and are deviated towards the production of androgens. In girl infants virilism appears as pseudo-hermaphroditism and in boys sexual precocity occurs. The condition is most often inherited as an autosomal recessive character and is usually apparent at birth or in infancy, but on occasions it may not be diagnosed until later in life. In many instances the enzyme deficiences are incomplete.

Deficiency of C-21 hydroxylase

The commonest form of congenital adrenal hyperplasia is due to a block in hydroxylation at C-21 (Fig. 7.1). An accumulation of the intermediate compound 17α-hydroxyprogesterone occurs, with a considerable increase in its urinary metabolites, pregnane-3α,17α,20α-triol, and 17-hydroxypregnanolone. Some 11-oxo-pregnanetriol is also formed and this may be greater in amount than pregnanetriol itself in very young infants. Urinary 17-oxosteroids are usually considerably raised above the normal infant value of less than 1 mg. per 24 hours. Unlike the relative absence of 11-oxy-C_{21} compounds, a larger proportion than normal of the total C_{19} steroids in the urine consists of 11-oxy-17-oxosteroids, especially 11β-hydroxyandrosterone, the adrenal source of this 5α-compound being 11β-androstenedione (see Chapter 4).

The diagnosis of C-21 hydroxylase deficiency is normally made by the finding of raised total 17-oxosteroids and pregnanetriol levels in the urine. Hill (1960) finds separation of the total 17-oxogenic steroids (17-hydroxycorticosteroids) into 11-deoxy and 11-oxy fractions to be a useful diagnostic procedure. In normal children the ratio of 11-deoxy to 11-oxy-17-oxogenic steroids is less than 0·5 whereas in severe forms of C-21 hydroxylase deficiency it is greater than 1·0. This method has the advantage that it can be carried out on a random sample of urine instead of waiting for a complete 24 hour specimen.

Deficiency of 11β-hydroxylase

An uncommon form of congenital adrenal hyperplasia with virilism occurs in which the defect lies in 11β-hydroxylation, so that there is an accumulation of the immediate precursors of cortisol and corticosterone,

11-deoxycortisol and 11-deoxycorticosterone (Fig. 7.2). The latter compound is thought to be responsible for the hypertension which is often a clinical feature of this type of congenital enzyme defect. Tetrahydro-11-deoxycortisol and tetrahydro-deoxycorticosterone are found in large quantities in the urine together with a moderate increase in pregnanetriol. Little or no 11-oxy-17-oxosteroids are present. An exactly comparable enzyme defect can be induced by the administration of the drug metyrapone.

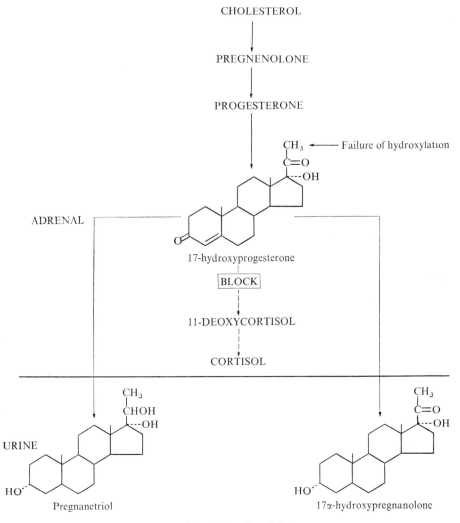

FIG. 7.1. C-21 hydroxylase deficiency.

Deficiency of 3β-hydroxysteroid dehydrogenase

In 1961 Bongiovanni described a rare form of congenital adrenal hyperplasia in which there was a deficiency of 3β-hydroxysteroid dehydrogenase (Fig. 7.3). The defect in cortisol synthesis occurs at an early stage, the conversion of pregnenolone to progesterone. Large amounts of the 3β-hydroxy-Δ^5-steroids are found in the urine, especially the C_{21} steroid pregn-5-ene-

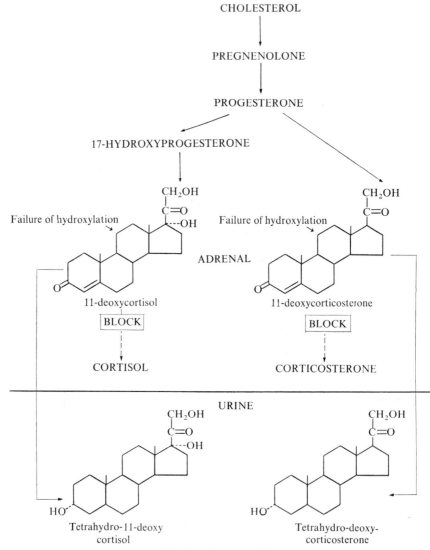

FIG. 7 2. 11β-hydroxylase deficiency.

$3\beta,17\alpha,20\alpha$-triol, and the C_{19} compound dehydroepiandrosterone. The latter is a relatively weak androgen and this may explain why virilization is less severe in this type of deficiency, since very little dehydroepiandrosterone is converted to testosterone in the peripheral tissues. For unexplained reasons, the mortality rate is very high in this form of defect.

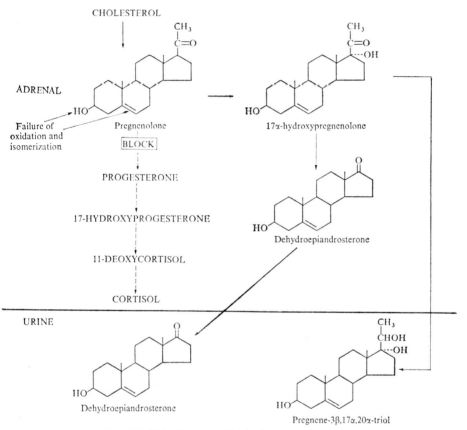

FIG. 7.3. 3β-hydroxysteroid dehydrogenase deficiency.

Deficiency of 17α-hydroxylase

Biglieri *et al.* (1966) described a female patient who had amenorrhoea, hypertension, and signs of mineralocorticoid excess which was due to an increased secretion of corticosterone, whereas aldosterone was decreased. No cortisol or other 17-hydroxylated compounds were present and there was a failure of sexual development in contrast with the virilism which is a feature of other forms of congenital enzyme defects. The lack of 17-hydroxylase seemed to involve both the adrenals and the ovaries since 17-oxosteroids and oestrogens were practically absent from the urine. The reduction in aldosterone secretion may have been related to the sodium retention caused by corticosterone excess, but could have been due to a separate enzyme defect in addition to that affecting 17α-hydroxylase. Other less severe cases of 17-hydroxylase deficiency have now been reported in which aldosterone secretion is maintained.

Salt-losing forms of congenital adrenal hyperplasia

In the most severe forms of C-21 hydroxylase deficiency and in the 3β-hydroxysteroid dehydrogenase defect, excessive urinary loss of sodium occurs which, unless recognized, rapidly leads to death of the infant. There has been much discussion as to the cause of the excessive sodium loss. The production of an actual salt-excreting hormone has been suggested by some workers but the existence of such a hormone has not been substantiated and careful estimation of aldosterone secretion in the severe forms of the C-21 hydroxylase defect shows it to be very low. It is probable that aldosterone deficiency is mainly, if not wholly, responsible for the severe sodium loss. Some workers claim that additional factors must be involved because of the comparatively large doses of a mineralcorticoid which are necessary to correct the salt-losing condition, but in fact the amount is no greater than that required for the treatment of Addison's disease in infancy and it seems that at this age the renal response to aldosterone is different from that of the adult. In addition to the reduction in aldosterone secretion which accompanies the severe form of C-21 hydroxylase deficiency, Ulick et al. (1964) have reported a patient who had an isolated failure of aldosterone biosynthesis. According to these authors, the enzyme defect in this patient was in the dehydrogenation of 18-hydroxycorticosterone to form aldosterone.

The androgens in congenital adrenal hyperplasia

The virilism which is a striking clinical feature of congenital adrenal hyperplasia is due to an excessive secretion of androgen, the source of which is the adrenal since suppression of androgen secretion and excessive 17-oxosteroid excretion can be achieved by the therapeutic use of cortisone or other glucocorticoids. The adrenal source of the C_{19} compounds found in the urine comprises dehydroepiandrosterone, androst-4-ene-3,17-dione, and 11β-hydroxyandrostenedione. These compounds are all rather weak androgens, but several workers have now reported that the concentration of the very potent androgen testosterone is increased in the peripheral blood of children with congenital adrenal hyperplasia. Horton & Frasier (1967) have shown that much of the increase in testosterone is derived from the conversion of androstenedione in extra-adrenal tissues. Congenital adrenal hyperplasia in its various forms is well reviewed by Bongiovanni et al. (1967).

Adrenal virilism in the adult female

Virilism due to the presence of an adrenocortical tumour is rare and is discussed in Chapter 9. There is, however, a comparatively large number of females who develop hirsutism after puberty with or without additional evidence of virilism and menstrual irregularities, the cause of which is not always easily determined. Most of these patients have normal or only moderately raised urinary 17-oxosteroid levels. A very small number have an increased excretion of pregnanetriol, and these are probably cases of mild C-21 hydroxylase deficiency which have escaped diagnosis until after puberty. In the great majority of patients pregnanetriol excretion is normal, but in spite of normal urinary 17-oxosteroid levels, the concentration of testosterone in the plasma is often increased above the normal female range of 0·025–0·068 μg per 100 ml. The excess androgen may be adrenal or ovarian in origin. The estimation of total 17-oxosteroids as routinely performed is of little assistance in the diagnosis but fractionation into the individual components may be be more useful. An excessive proportion of dehydroepiandrosterone which is readily suppressed by the administration of dexamethasone and which is disproportionately increased after giving ACTH, favours a non-tumorous adrenal origin. If dehydroepiandrosterone is normal in amount but androsterone and aetiocholanolone are relatively increased and if they are only slightly reduced by dexamethasone the androgen is more likely to originate from the ovary. Rarely, an ovarian tumour is responsible, particularly in those cases of post-pubertal virilism of fairly rapid onset and a normal 17-oxosteroid excretion. Much more commonly, there is hirsutism, or only slight evidence of virilism together with irregular menstruation which are associated with bilaterally enlarged cystic ovaries. The condition then constitutes the Stein–Leventhal syndrome. There is ample evidence that these cystic ovaries can secrete excessive amounts of androgen and can be responsible for the increased level of plasma testosterone which is often found. Stimulation of the ovaries by chorionic gonadotrophin, or, according to some workers, follicle stimulating hormone, results in even greater amounts of androgen. Undoubtedly, however, there is a proportion of patients with the polycystic ovary syndrome whose urinary 17-oxosteroids and plasma testosterone are suppressed by the administration of dexamethasone. In these patients the androgen is presumed to be of adrenal origin but the relationship between the ovary and the adrenal cortex in the development of virilism is still far from clear. Virilism in the post-pubertal female is reviewed in detail by Prunty (1967).

CHEMICAL INHIBITORS OF ADRENOCORTICAL SECRETION

Many chemically diverse substance are now known which, in various ways,

E

can inhibit adrenal steroid synthesis. Some of these substances act by causing necrosis of adrenal tissue but others inhibit adrenal enzyme systems, either generally or selectively (Fig. 7.4).

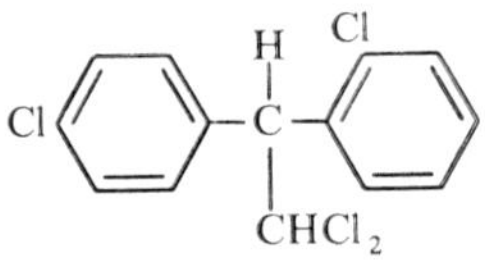

2,2-bis(4-chlorophenyl,2-chlorophenyl)-
1,1-dichloroethane

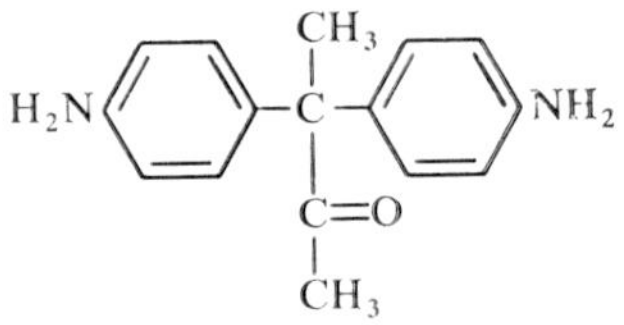

DDD

3,3-bis(p-aminophenyl)-2-butanone.

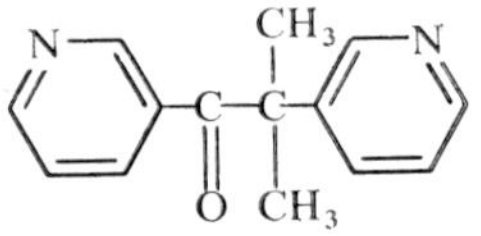

Amphenone

2-methyl-1,2-bis(3-pyridyl)-1-propanone

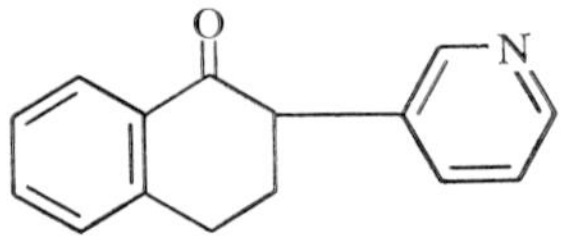

Metyrapone

3-(1,2,3,4-tetrahydro-1-oxo-2-naphthyl)-
pyridine

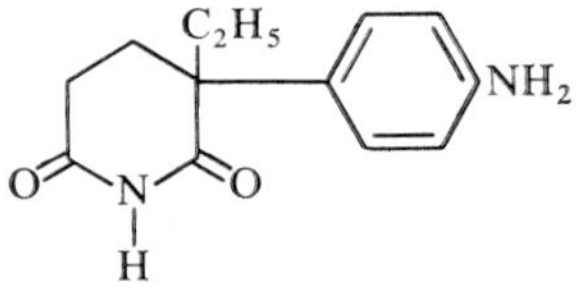

SU 9055

α-ethyl-α-p-aminophenylglutarimide

Aminoglutethimide

Fig. 7.4. Inhibitors of adrenocortical secretion.

The most important of the agents which cause adrenal cortical necrosis is *o,p'* DDD, 2,2-bis(4-chlorophenyl,2-chlorophenyl)-1,1-dichloroethane, a compound which is closely similar to the insecticide DDT. By reason of its destructive action it causes a generalized reduction in adrenal steroid secretion and is used in the treatment of adrenal carcinoma, but its toxic effects severely limit its usefulness. It has also been found that DDD alters the metabolism of cortisol in such a way that more 6β-hydroxycortisol and less tetrahydro derivatives are produced with the result that the urinary 17-hydroxycorticosteroids, as usually estimated, are greatly decreased. The urinary steroid values are not therefore an accurate guide to depression of cortisol synthesis, which is best determined by the levels of plasma cortisol or its actual secretion rate.

Other agents which cause necrosis of the adrenal cortex experimentally are known, such as the anticoagulant compound hexadimethrine, and the aromatic hydrocarbon 7,12-dimethylbenzanthracene, but they have not found any use in man.

Inhibitors of adrenocortical enzymes

The first adrenal inhibitor to be used was amphenone B, 3,3-bis (*p*-aminophenyl)-2-butanone, which, in contrast with DDD, produces an increase in adrenal weight if given for a prolonged period to experimental animals. The enlargement of the adrenals is due to an increase in the release of ACTH which follows inhibition of cortisol secretion. The inhibition takes place at multiple sites in the synthesis of cortisol from the early to the last steps in the sequence of enzyme reactions. Amphenone has been used in the treatment of Cushing's syndrome and adrenal carcinoma but toxic effects are too severe to justify clinical use.

Metyrapone ("*Metopirone*", *SU* 4885)

The most widely used of the adrenal inhibitors is an analogue of amphenone known as metyrapone, which is 2-methyl-1,2-bis-(3-pyridyl)-1-propanone. Jenkins *et al.* (1958) and Liddle *et al.* (1958) independently discovered that this substance selectively inhibited 11β-hydroxylation when given in the appropriate dosage to the dog and man. After the administration of metyrapone, the immediate precursors of cortisol and corticosterone, 11-deoxycortisol and 11-deoxycorticosterone are formed by the adrenal and corresponding tetrahydro-derivatives are found in the urine. The effect is exactly comparable to the naturally occurring congenital defect in 11β-hydroxylation (Fig. 7.2). The degree of 11β-hydroxylase inhibition is dependent upon the dose used. The inhibition of cortisol secretion results in an increased release of ACTH, with a further increase in the formation of 11-deoxy compounds including urinary tetrahydro-11-deoxycortisol, and this is measured by the usual techniques for the estimation of 17-oxogenic steroids or 17-hydroxycorticosteroids (see Chapter 10). Although inhibition of 11β-hydroxylation also leads to a reduction in aldosterone secretion, the effect of this is offset by the formation of 11-deoxycorticosterone, a potent mineralocorticoid. Inhibition of aldosterone without interference from an

E*

excess of deoxycorticosterone can be achieved by the simultaneous administration of prednisone to suppress release of ACTH.

By far the most important use of metyrapone is in the assessment of
hypothalamic-pituitary-adrenal function, since the increase in urinary
17-hydroxycorticosteroids which it produces in the normal individual
provides a measure of the integrity of the feedback control of ACTH release
(see Chapter 12).

Attempts have been made to develop other specific inhibitors of cortisol
synthesis. The compound, 3-(1,2,3,4-tetrahydro-1-oxo-2-naphthyl)-pyridine,
(SU 9055) has been reported by Chart et al. (1962) to be an inhibitor of
17α-hydroxylase, so that this substance causes a reduction in the secretion of
the 17-hydroxylated steroids such as cortisol, and an increase in corticosterone.

Inhibitors of aldosterone secretion

From the therapeutic point of view it is sometimes desirable to reduce the
effects of aldosterone secretion. Most of the inhibitors of cortisol already
described will also inhibit aldosterone synthesis but they are either too
toxic, or, in the case of metyrapone, require the simultaneous suppression of
ACTH secretion by the administration of a glucocorticoid. It is of interest
that although aldosterone is not a 17α-hydroxylated steroid, the 17α-hydroxy-
lase inhibitor SU 9055 also inhibits aldosterone secretion, probably by a
separate inhibitory effect on 18-hydroxy dehydrogenase (Bledsoe et al.
1964).

The sodium-excreting action of prolonged heparin administration led to a
study of the effect of heparin and its analogues on aldosterone secretion.
The most studied of these "heparinoids" is R01-8307, (N-formyl-chitosan
polysulphuric acid), which has very weak anticoagulant properties but considerably reduces the adrenal biosynthesis of aldosterone without any effect
on cortisol secretion (Conn et al. 1966).

The compound most often used therapeutically to counteract excessive
aldosterone secretion is not, however, an inhibitor of aldosterone synthesis
but is an antagonist of the hormone at its site of action on the renal tubule.
This substance is a synthetic steroid lactone, spironolactone, (Aldactone)
(Fig. 7.5), and in fact owing to the sodium excretion which it causes, spironolactone provokes an increase in the secretion of aldosterone.

Other inhibitors of adrenocortical enzymes

Apart from deliberate attempts to seek adrenal inhibitors it is of some
interest and importance that drugs which are already in use for quite different
purposes may be found by chance to be adrenal inhibitors. Thus, the anti-
convulsant drug, aminoglutethimide, has been found to reduce the synthesis
of cortisol and aldosterone in man by inhibiting adrenal steroid synthesis

from cholesterol at the very earliest stage, the conversion of cholesterol to pregnenolone. The effect on cortisol synthesis, however, is limited by the compensatory increase in release of ACTH which comes into play, and which restores cortisol secretion to normal except in those cases of Cushing's syndrome which are outside pituitary control. The inhibitory effect on

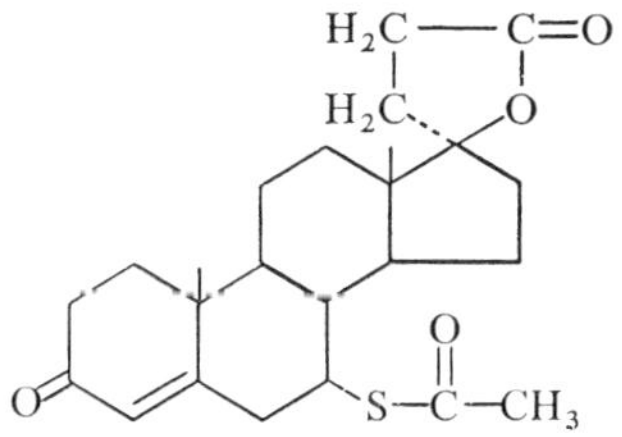

FIG. 7.5. Spironolactone.

aldosterone secretion, however, remains unaffected (Fishman *et al.* 1967). Another example of an unsuspected adrenal inhibitory effect is provided by the anti-depressant drug tranylcypromine. Johnson *et al.* (1967) have reported that in dogs this substance inhibits 17α-hydroxylation but in doses greater than are usually used in man. It seems likely that other drugs will be found to be adrenocortical inhibitors.

REFERENCES

BIGLIERI, E. G., HERRON, M. A., and BRUST, N. (1966). *J. clin. Invest.* **45**, 1946–1945.
BLEDSOE, T., ISLAND, D. P., RIONDEL, A. M., and LIDDLE, G. W. (1964). *J. clin. Endocr.* **24**, 740–746.
BONGIOVANNI, A. M., EBERLEIN, W. R., GOLDMAN, A. S., and NEW, M. (1967). *Recent Progr. Hormone Res.* **23**, 375–449.
CHART, J. J., SHEPPARD, H., MOWLES, T., and HOWIE, N. (1962). *Endocrinology,* **71**, 479–486.
CONN, J. W., ROVNER, D. R., COHEN, E. L., and ANDERSON, J. E. (1966). *J. clin. Endocr.* **26**, 527–532.
FISHMAN, L. M., LIDDLE, G. W., ISLAND, D. P., FLEISCHER, N., and KUCHEL, O. (1967). *J. clin. Endocr.* **27**, 481–490.
HILL, E. E. (1960). *Acta Endocr.* (Kbh). **33**, 230–250.
HORTON, R. and FRASIER, S. D. (1967). *J. clin. Invest.* **46**, 1003–1009.
JENKINS, J. S., MEAKIN, J. W., NELSON, D. H., and THORN, G. W. (1958). *Science,* **128**, 478–480.
JOHNSON, P. C., LORENZEN, L. C., BIGLIERI, E. G., and GANONG, W. F. (1967). *Endocrinology.* **80**, 510–514.
LIDDLE, G. W., ISLAND, D., LANCE, E. M., and HARRIS, A. P. (1958). *J. clin. Endocr.* **18**, 906–912.
POSNER, J. B. and JACOBS, D. R. (1964). *Metabolism,* **13**, 513–521.
PRUNTY, F. T. G. (1967). *J. Endocr.* **38**, 85–103, 203–227.
ULICK, S., GAUTIER, E., VETTER, K. K., MARKELLO, J. R., YAFFE, S., and LOWE, C. U. (1964). *J. clin. Endocr.* **24**, 669–672.
VANCE, V. K., REDDY, W. J., NELSON, D. H., and THORN, G. W. (1962). *J. clin. Invest.* **41**, 20–28.

8

DISORDERS OF METABOLISM OF THE ADRENAL STEROIDS

The normal metabolism of the adrenal cortical hormones and the enzyme reactions concerned are discussed in Chapter 4. Diseases affecting the tissues where these chemical changes take place influence both the pattern of the metabolites formed and the rate of their formation.

LIVER DISEASE

Owing to the important role of the liver in adrenal steroid metabolism, considerable disturbances in the chemical transformation of cortisol occur when this organ is diseased. In advanced cirrhosis of the liver, for example, the levels of urinary 17-hydroxycorticosteroids are below normal, mainly due to decreased amounts of the tetrahydro-derivatives of cortisol and cortisone. The decreased metabolism of cortisol in cirrhosis is seen even more clearly in studies on the disappearance rate of injected hormone, in which the half-life time may greatly exceed the normal value (Table 8.1).

TABLE 8.1. Effect of hepatic and thyroid disease on cortisol metabolism

Diagnosis	*Cortisol half-life (min)	Plasma 17-hydroxy-corticosteroids (μg/100 ml.)
Hepatic cirrhosis	162	17
Hepatic cirrhosis	192	23
Hyperthyroid	69	16
Hyperthyroid	66	15
Hyperthyroid	66	14
Hypothyroid	141	11
Hypothyroid	144	14
Hypothyroid	150	6
Normal range	102–114	6–25

* The half-life times are derived from the disappearance of cortisol from the blood after the intravenous administration of 100 mg of the steroid. When tracer doses are used the normal half-time is shorter and is about 80 minutes.

In spite of its diminished breakdown, the concentration of cortisol in the plasma remains within the normal range. This has been shown by Peterson (1960) and others to be due to a fall in the cortisol secretion rate, which parallels the reduction in metabolism. Not all the liver enzymes taking part in the chemical transformation of cortisol are equally affected in disease of the liver. The main defect is found in the Δ^4-reductase enzymes which reduce the C-4,5 double bond to form 5α- or 5β-dihydrocortisol, and it is the limitation of this step which is responsible for the decreased amounts of conjugated tetrahydro compounds found in the urine. Peterson (1960) has shown that the further reduction of dihydrocortisol at the 3-oxo group to form tetrahydrocortisol and its subsequent conjugation with glucuronic acid are not significantly impaired in cirrhosis of the liver since the half-life times of injected dihydrocortisol and tetrahydrocortisol are normal. Jenkins & Sampson (1967) found that the 11β-hydroxydehydrogenase enzyme, which is responsible for the reduction of cortisone to cortisol, or prednisone to prednisolone, was not obviously impaired in severe infectious hepatitis or in cirrhosis, whereas reduction of ring A of both cortisone and cortisol was greatly affected. The cortisol metabolite 6β-hydroxycortisol is increased in liver disease, and this occurs in many conditions where reduction of ring A is deficient.

Aldosterone and liver disease

In contrast with the decreased secretion rate of cortisol, aldosterone secretion is often increased in cirrhosis of the liver, especially when ascites is present. The mechanism involved in the increased aldosterone secretion found in ascites and other oedematous states is not certain. A reduction in the plasma volume may be a factor together with sodium depletion induced therapeutically. Again, in contradistinction to the normal plasma concentration of cortisol, the level of aldosterone in the blood is often raised, sometimes to 10 or 15 times the normal values. Metabolism of aldosterone is, however, decreased so that the half-life time may be doubled from the normal of about 33 minutes. There is a reduction in the clearance of aldosterone by the diseased liver and the formation of tetrahydroaldosterone is impaired. There is, however, a relative increase in the proportion of the acid-labile conjugate present in the urine. 50% of this derivative of aldosterone is known to be formed outside the liver, within the kidney, and in hepatic cirrhosis a greater amount than normal of unmetabolized aldosterone is presented to the kidney (Bledsoe *et al.* 1966).

RENAL DISEASE

In man the excretion of adrenal steroid metabolites, as their conjugates, takes place almost entirely through the kidney so that impairment of renal function is likely to lead to a retention of conjugated steroids in the blood. Englert *et al.* (1958) showed that in uraemia the levels of urinary 17-hydroxy-corticosteroids were greatly reduced and they were correlated with the degree of renal insufficiency, as measured by the creatinine clearance. The conjugated metabolites of cortisol in the plasma, consisting mainly of the glucuronides of tetrahydrocortisol and tetrahydrocortisone were elevated in all cases of uraemia, although the concentration of unconjugated 17-hydroxycorticosteroids was within the normal range. The infusion of cortisol in these patients showed, somewhat unexpectedly, that the half-life time was prolonged to about 50% above normal. The reason for this finding is not clear, but it should be remembered that in addition to its excretory function the human kidney has the capability, at least *in vitro*, of metabolizing cortisol to cortisone and 20-dihydrocortisol. The importance of the metabolic role of the kidney in man is still not known, but it is possible that the development of uraemia may result in some decrease in cortisol metabolism in addition to the more obvious failure to excrete preformed metabolites.

The kidney and aldosterone

The relationship of renal function to aldosterone metabolism is more complicated than that of cortisol, since one of the major regulating mechanisms in the control of aldosterone secretion is renin, which is itself a product of the kidney. In the presence of either unilateral or bilateral kidney disease, the renal juxtaglomerular apparatus liberates an increased amount of renin, which then stimulates the synthesis of aldosterone. The raised aldosterone secretion found in many cases of malignant hypertension is correlated with the elevated renin levels present in this disease, whereas in the benign form of hypertension, the concentrations of both aldosterone and renin are usually normal. One of the earliest observations relating to aldosterone production was the finding of increased amounts in the urine of some patients suffering from nephrosis. In this form of renal disease there is marked peripheral oedema, and the resultant fall in plasma volume may also stimulate the renin mechanism. Renal failure causes a decrease in the excretion of tetrahydro-aldosterone glucuronide and the acid-labile glucuronide, part of which is formed in the kidney. For this reason, in the presence of uraemia estimations of the aldosterone secretion rate which are based on the estimation of a urinary metabolite may be fallacious.

DISORDERS OF THE THYROID

It is now well established that the thyroid gland influences adrenal steroid metabolism. In thyrotoxicosis there is an increased turnover of cortisol, and a shortening of the half-life time (Table 8.1), but the plasma levels of the hormone remain within the normal range, since there is a compensatory increase in the cortisol secretion rate (Peterson, 1958). Conversely, in myxoedema there is a decreased rate of metabolism and a decrease in the cortisol secretion rate, the levels in the plasma again being normal (Table 8.1). By reason of the feedback control, the thyroid, through its action on the liver enzymes, exerts a regulating influence on cortisol secretion. Administration of thyroxine to rats causes an increase in the reduction of ring A of cortisol probably by increasing both the Δ^4 reductase enzymes in the liver and also the amount of available $NADPH_2$. In thyrotoxicosis there is a moderate increase in the levels of urinary 17-hydroxycorticosteroids and in myxoedema there is a decrease. These changes revert to normal when the thyroid state is treated appropriately. In addition to these quantitative aspects of cortisol metabolism, changes in the pattern of metabolites are also found. Hellman *et al.* (1961) showed that in thyrotoxicosis there is an increase in the oxidation of the 11β-hydroxyl group. This results in greater than normal amounts of the 11-oxo compounds tetrahydrocortisone, 11-oxo aetiocholanolone, and cortolone being present in the urine. In myxoedema there is a preponderance of 11β-hydroxysteroids, so that there is a greater amount of urinary tetrahydrocortisol than tetrahydrocortisone, which is a reversal of the normal finding. In the case of the urinary 17-oxosteroids, there are significant alterations in the proportion of the 5α and 5β-isomers androsterone and aetiocholanolone. In thyrotoxicosis, or after the administration of thyroxine, there is an increase in the amount of the 5α-steroid, androsterone. It is of interest that the administration of androsterone, like thyroxine, leads to a fall in the serum cholesterol.

Disorders of extra-hepatic metabolism

The fact that metabolism of adrenal steroids can take place outside the liver is well substantiated, and there are several reports of alterations in metabolism which accompany diseases of extra-hepatic tissues. The possibility that renal disease may affect cortisol metabolism has already been discussed. Berliner & Dougherty (1961) have reviewed the evidence that some types of malignant cells metabolize cortisol to a greater extent than their non-malignant counterparts. It has been reported that patients with acute lymphoblastic leukaemia excrete more 11β-hydroxyaetiocholanolone than normal subjects, and the writer has found that leukaemic lymphoblasts have a greater capacity to metabolize cortisol *in vitro* than normal "blast" cells. It has been suggested that one result of the increased metabolism of cortisol by leukaemic cells is to render those cells more resistant to the usual inhibitory action of the hormone. Normal prostatic tissue slices can oxidize the α-ketol side chain of cortisol to a C_{19} steroid *in vitro* to a limited extent, and it has been found that patients with carcinoma of the prostate convert cortisone and cortisol to 17-oxosteroids in greater amounts than normal. The relative importance, however, of the changes in steroid metabolism seen in disease processes affecting peripheral tissues remains to be determined.

The effect of drugs on adrenal steroid metabolism

Several drugs and some steroid hormones are known to affect the metabolism of cortisol. The administration of oestrogens is followed within 3 to 7 days by an increase in the protein-binding of cortisol by corticosteroid-binding globulin, which may protect cortisol from inactivation by the liver since there is a subsequent decrease in the formation of tetrahydro derivatives and a prolongation of the half-life time of cortisol. In addition, there is an increased formation of 6β-hydroxycortisol but this increase does not seem to depend on alterations in protein-binding since it can take place when liver slices are exposed to oestrogen *in vitro* (Lipman *et al.* 1962).

Certain inhibitors of adrenal steroid synthesis such as aminoglutethamide and *o-p* DDD affect cortisol metabolism by decreasing the formation of tetrahydro compounds and, in the case of DDD, an increase in 6β-hydroxycortisol has been demonstrated. The importance of this finding is that estimation of urinary 17-hydroxycorticosteroids alone cannot provide a reliable guide to the degree of actual adrenal cortical inhibition obtained by the use of these agents.

Other examples of drugs affecting cortisol metabolism are the anticonvulsants phenobarbitone and phenytoin, and the anti-inflammatory substance phenylbutazone. These drugs decrease ring A reduction but increase the secretion rate of 6β-hydroxycortisol when given for a period of several days, probably by the induction of increased amounts of the enzyme 6β hydroxylase within the liver (Burstein *et al.* 1967). Such enzyme induction by phenobarbitone is well substantiated in other metabolic reactions (Yaffe *et al.* 1966).

REFERENCES

BERLINER, D. L. and DOUGHERTY, T. F. (1961). *Pharmacol. Rev.* **13**, 329–359.
BLEDSOE, T., LIDDLE, G. W., RIONDEL, A., ISLAND, D. P., BLOOMFIELD, D., and SINCLAIR-SMITH, B. (1966). *J. clin. Invest.* **45**, 264–269.
BURSTEIN, S., KIMBALL, H. L., KALIBER, E. L., and GUT, M. (1967). *J. clin. Endocr.* **27**, 491–499.
ENGLERT, E., DROWN, H., WILLARDSON, D. G., WALLACH, S., and SIMONS, E. L. (1958). *J. clin. Endocr.* **18**, 36–48.
HELLMAN, L., BRADLOW, H. L., ZUMOFF, B., and GALLAGHER, T. F. (1961). *J. clin. Endocr.* **21**, 1231–1247.
JENKINS, J. S. and SAMPSON, P. A. (1967). *Brit, Med. J.* **2**, 205–207.
LIPMAN, M. M., KATZ, F. H., and JAILER, J. W. (1962). *J. clin. Endocr.* **22**, 268–272.
PETERSON, R. E. (1958). *J. clin. Invest.* **37**, 736–743.
PETERSON, R. E. (1960). *J. clin. Invest.* **39**, 320–331.
YAFFE, S. D., LEVY, G., MATSUZAWA, T., and BALIAH, T. (1966). *New Eng. J. Med.* **275**, 1461–1466.

9

DISORDERS OF CONTROL OF
ADRENOCORTICAL SECRETION

Abnormalities of the mechanisms responsible for the control of adrenocortical secretion may result in either decreased or increased hormone production.

HYPOPITUITARISM

Lesions of the anterior part of the pituitary, if sufficiently severe, result in an impaired release of ACTH and a fall in adrenocortical secretion. The deficiency may be apparent under basal conditions or only after exposure of the patient to a stressful stimulus. In severe hypopituitarism the level of cortisol and corticosterone in the blood may approach zero. The secretion of aldosterone, however, is often maintained at a fairly normal level, since this hormone is mostly under extra-pituitary control. Disturbances of electrolyte metabolism are not therefore present to the same degree in hypopituitarism as in Addison's disease. The values for urinary 17-oxogenic steroids are greatly reduced in severe anterior pituitary insufficiency and the urinary 17-oxosteroids are extremely low in both sexes since there is usually an associated loss of gonadal function. The administration of ACTH evokes some response from the adrenal and differentiates the condition from the primary adrenal insufficiency of Addison's disease, but the rise in cortisol secretion is often sluggish and treatment with ACTH for several days may be required before a response is observed. It must be emphasized, however, that an ACTH test measures only the adrenal responsiveness and gives no indication of the ability of the pituitary to release endogenous ACTH. This factor is particularly important to assess in those patients whose basal cortisol levels are within the normal range but whose ability to react to stress may be in doubt. Several tests have now been devised which in various ways measure different aspects of pituitary-adrenal function. Some, such as the metyrapone test, assess the feedback control, while others, such as pyrogen and insulin hypoglycaemia probably reflect the response to stress and are discussed in detail in Chapter 12. Loss of pituitary function may be incomplete and secretion of gonadotrophin and growth hormone are frequently lost before that of ACTH; very rarely isolated ACTH deficiency occurs.

DISEASES OF THE CENTRAL NERVOUS SYSTEM

The control of adrenocortical function which the central nervous system exerts through the hypothalamus and pituitary (Chapter 5) may become disturbed in various disorders affecting the brain. Tumours of the hypothalmic region particularly, and inflammatory or vascular lesions involving this area may lead to a clinical and biochemical picture similar to that seen in lesions of the pituitary itself. Damage to the hypothalamus may interfere with the secretion of corticotrophin-releasing factor together with the other hypothalamic factors responsible for the release of gonadotrophins, thyrotrophin, and growth hormone from the anterior pituitary. Variation in the degree to which the different releasing factors are affected, leads to differences in the picture of hypopituitarism produced, and in some cases only one or two deficiencies may be apparent. In lesions of the anterior part of the hypothalamus affecting the supra-optic area, there is, in addition, loss of vasopressin secretion, resulting in diabetes insipidus. This condition may, of course, occur alone without any apparent loss of hypothalamic-anterior pituitary function, when this is assessed by the methods described in Chapter 12.

Certain drugs active upon the central nervous system may interfere with the hypothalamic pathways. Thus chlorpromazine and the anti-convulsant drug phenytoin have been found to inhibit the normal response to metyrapone but not to other tests of hypothalamic-pituitary-adrenal function such as pyrogen and vasopressin. The latter is, however, inhibited by morphine (Jenkins, 1968).

Inhibition of pituitary-adrenal function by corticosteroid therapy

One of the most serious side-effects of glucocorticoid therapy, when administered in a dosage greater than the physiological cortisol secretion of about 20 mg daily, is adrenocortical suppression. A state of hypopituitarism is induced, as far as the release of ACTH is concerned, by the action of the steroid on the feedback control within the hypothalamus to inhibit ACTH release. It can be shown that after the administration of a single large dose of the synthetic glucocorticoid dexamethasone, which does not interfere with the fluorimetric estimation of cortisol, plasma levels of cortisol fall to very low values within two hours, but the levels return to normal within 24 hours. Knowledge of the length of time that the adrenal remains suppressed after the cessation of prolonged steroid therapy, is most important, and while it depends to some extent upon the duration of therapy, there is great variation between different individual patients. Two factors are involved; the recovery of the pituitary ACTH-releasing mechanism and the responsiveness of the adrenal cortex to ACTH. Careful studies by Graber *et al.* (1965) who have measured directly the plasma concentration of both ACTH and cortisol, indicate that the pituitary usually recovers first. Patients who have received prolonged corticosteroid therapy may be at risk when exposed to a stressful situation such as surgery and it would be very useful, therefore, to be able to predict the hypothalamic-pituitary-adrenal response in such patients. In the opinion of the writer, the application of the pyrogen, or insulin hypoglycaemia tests (Farmer *et al.* 1961; Livanou *et al.* 1967) are particularly useful in this respect, whereas the more often used metyrapone tests is of less value since it is a measure of the feedback control rather than the response to stress (see Chapter 12).

CUSHING'S SYNDROME

The essential basis of Cushing's syndrome is an excessive secretion of cortisol, whereas aldosterone production is almost always normal. There is sometimes an excessive secretion of adrenal androgens and, rarely, oestrogens. Androgen secretion is very variable and ranges from slight in the case of a non-virilizing adrenal adenoma, to a great excess in many cases of adrenal cortical carcinoma. When the clinical picture is dominated by virilism the condition is then designated the adrenogenital syndrome. The increased activity of the adrenal cortex which is responsible for the excessive secretion of cortisol in Cushing's syndrome may have one of four possible causes; bilateral adrenal hyperplasia, benign adenoma of one adrenal cortex, carcinoma of the adrenal cortex, or extra-adrenal malignant disease.

Bilateral adrenal hyperplasia

In over 75% of most large series of patients with Cushing's syndrome, bilateral hyperplasia of the adrenal cortex is present. The cause of this adrenal enlargement and over-activity has been the subject of some discussion, but most authorities have favoured the view that the anterior pituitary is the primary site of origin. It is only in recent years, however, with the development of very sensitive techniques for the estimation of plasma levels of ACTH that it has been possible to demonstrate directly that that there is an increased secretion of ACTH in this form of Cushing's syndrome. It has been found that ACTH levels are not always greatly elevated above normal but an explanation for the increased adrenal activity is that the normal diurnal variation in ACTH release is lost, so that over the whole period of 24 hours the secretion of ACTH is excessive. The relative resistance of patients with

Cushing's syndrome to pituitary-adrenal suppression after the administration of glucocorticoids is further evidence that the ACTH-secreting mechanism is set at a higher level than normal, and provides a useful diagnostic procedure. In some patients with bilateral adrenal hyperplasia the surgical removal of the adrenals is followed by the development of a demonstable pituitary adenoma which secretes very large amounts of ACTH and MSH, and leads to a rapid increase in pigmentation of the skin (Nelson *et al.* 1960). Urinary 17-oxogenic steroids may be elevated in Cushing's syndrome but in up to 50% of cases they are within the upper normal range so that they are not necessarily of diagnostic value. Plasma cortisol levels are often raised, but again, fall within the normal range in an appreciable number of patients. Of much more value is the estimation of plasma cortisol in the morning and at night, since a loss of the normal diurnal variation is present in most cases. Chromatographic analysis of the urinary metabolites of cortisol shows a greater proportion of tetrahydrocortisol than tetrahydrocortisone, which is the reverse of the normal findings, and a greater production of the very polar metabolite, 6β-hydroxycortisol. Much more unmetabolized cortisol than normal is also present in the urine and some workers have found this to be a useful aid in diagnosis.

The estimation of 17-oxosteroids is not of specific diagnostic value in hyperplasia but there is an increased proportion of 11-oxo-17-oxosteroids derived from the metabolism of cortisol.

Adrenocortical adenoma

Although many adenomata present in the adrenal cortex are small and non-functioning, in about 15% of cases of Cushing's syndrome, an adreno-cortical adenoma is found to be the cause, the opposite adrenal often being atrophic. The biochemical distinction from hyperplasia is not always well defined, but the finding of low 17-oxosteroid excretion relative to the raised values for 17-oxogenic steroids may sometimes give an indication of the presence of an adenoma. The response to suppression with dexamethasone or stimulation with ACTH is variable; some adenomata react like adrenal hyperplasia, others behave autonomously.

Adrenocortical carcinoma

The least common cause of Cushing' syndrome is a carcinoma of the adrenal cortex. In the very uncommon event of Cushing's syndrome occurring in children, however, a malignant origin is more likely. In addition to excessive secretion of cortisol, the tumour often secretes very large amounts of androgens leading to severe virilization. Urinary 17-oxosteroids sometimes amount to several hundred mg. per 24 hours and when values in excess of 50 mg per day are obtained, an adrenal carcinoma should be suspected. The androgens usually show a higher proportion than normal of dehydro-epiandrosterone and this indicates a relative lack of the 3β-hydroxysteroid dehydrogenase enzyme in the carcinomatous adrenal tissue. Cortisol synthesis may not only be excessive but may also be abnormal since large amounts of tetrahydro-11-deoxycortisol are sometimes found, indicating some failure of 11β-hydroxylation. Patients with these tumours may suffer from hypoglycaemia, an association which is not easy to explain (Williams *et al.* 1961). Very rarely, feminization occurs as a result of adrenocortical carcinoma and excessive amounts of oestrone and oestriol have been isolated from the urine of such cases. Adrenocortical carcinoma is a rare condition and Hutter & Kayhoe (1966) provide a useful review of a large number of cases.

Extra-adrenal malignant disease

The occasional association of Cushing's syndrome with extra-adrenal malignant disease, most often anaplastic carcinoma of the bronchus, has been known for many years. It is, however, only relatively recently that the biochemical basis of the condition has been elucidated. It has also become apparent that many cases do not present as classical Cushing's syndrome but as muscular weakness associated with hypokalaemia. The low serum potassium is not due to excessive aldosterone secretion but to an enormous increase in the production of cortisol so that secretion rates of over 400 mg per day have been recorded. The levels of plasma cortisol and urinary 17-oxogenic steroids are usually much higher than in Cushing's syndrome due to adrenal hyperplasia, in which marked hypokalaemia is only exceptionally found. The high cortisol level is unaffected by any methods of suppression or stimulation. Less commonly than bronchial carcinoma, malignant tumours of the thymus, breast, ovary, and pancreas are encountered and the patient may succumb to the malignant disease before there is time for the clinical features of Cushing's syndrome to develop. Liddle *et al.* (1965) have designated this condition "the ectopic ACTH syndrome" since by a wide variety of criteria, including biological, immunological and chromatographic properties, a substance closely similar to pituitary ACTH has been obtained from the tumours. MSH-like material has also been isolated and is presumably responsible for the increased pigmentation of the skin which is seen in many cases. The production of ACTH by carcinomatous tissue is not a unique process since other polypeptide hormones in association with tumours have now been recognized, including substances with the properties of parathormone, thyrotrophin, and anti-diuretic hormone. The nature of their synthesis is unknown but it would seem that certain malignant tumours have some degree of genetic derepression which allows the cells to manufacture polypeptides normally confined to specific endocrine gland tissue. It is also possible that the phenomenon of unrestrained polypeptide synthesis may be more general than hitherto suspected, and that the ectopic ACTH and other endocrine syndromes may be impressive biological examples of a widespread facility of malignant tissue.

The excessive production of cortisol responsible for the condition can be demonstrated most directly by estimation of the daily cortisol secretion rate, although many laboratories do not have the facilities for carrying out this procedure (see Chapter 14). Cope & Pearson (1965) found that in adreno-cortical hyperplasia cortisol production ranged from 36 mg to 138 mg daily and in adrenal carcinoma a value of 316 mg was obtained compared with a mean normal value of 16·2 mg. While a cortisol secretion rate of less than 20 mg daily is of value in the exclusion of Cushing's syndrome, a moderately raised secretion is sometimes found in simple obesity. A raised value for urinary 17-oxogenic steroids, even if serial determinations are carried out for several days, is found in only about 50% of cases of hyperplasia. Plasma cortisol estimations are of somewhat greater value, but the loss of the diurnal variation in plasma cortisol has a much higher diagnostic reliability, although it is also lost in some disorders of the central nervous system. Values at midnight are nearly always less than 10μg/100 ml. in normal individuals. A most useful diagnostic procedure and one which will also give some information as to the pathogenesis of the Cushing's syndrome is the dexamethasone suppression test introduced by Liddle. The administration of 2 mg dexamethasone daily in 4 doses of 0·5 mg for two days normally results in the urinary 17-oxogenic steroids falling to less than 5 mg/24 hours and the plasma cortisol to less than 5 μg/100 ml. If normal suppression does not occur, thus suggesting the presence of Cushing's syndrome, the dosage of dexamethasone is then raised to 8 mg daily for a further 2 days. Most cases of adrenal hyperplasia and some cases of adenoma will now respond with a decrease in urinary 17-oxogenic steroids of at least 50% but nearly all cases of adrenal carcinoma and those due to extra-adrenal malignancy will not suppress even after this large dose. Stimulation of the adrenal cortex with ACTH has also been used as a diagnostic procedure by some workers but there are difficulties in interpretation owing to the wide range of the normal response. A hyperactive response is obtained in adrenal hyperplasia, whereas most, though not all, cases of carcinoma fail to react to ACTH stimulation. Determination of daily urinary cortisol is of value in diagnosis since in Cushing's syndrome it is almost invariably greater than 90 μg/24 hours, but it requires a careful chromatographic technique and the simpler procedures outlined above are usually sufficient. The level of urinary 17-oxosteroids is of no diagnostic value in cases of adrenal hyperplasia where it is often normal or only slightly elevated. In adrenal carcinoma, however, very high levels may give an indication of the diagnosis.

Obesity

While Cushing's syndrome is a relatively uncommon condition, many patients with obesity and perhaps pink skin striations, diabetes or hypertension require investigation for biochemical evidence of Cushing's syndrome. Some cases of simple obesity have a moderately increased cortisol secretion rate and raised levels of urinary oxogenic steroids which cause diagnostic confusion (Migeon *et al.* 1963). The reason for these findings is not clear, but it has also been observed that obese patients metabolize cortisol at a higher rate than normal, so that the plasma levels are within the normal range. The normal diurnal variation is preserved, which provides an important diagnostic distinction from most cases of Cushing's syndrome. In addition, the dexamethasone suppression test is normal in patients suffering from obesity. Of interest is the finding that even with modest weight reduction, the cortisol secretion rate and urinary 17-oxogenic steroids fall to within the normal range.

A useful review of Cushing's syndrome and its diagnosis is provided by Ross *et al.* (1966).

PRIMARY HYPERALDOSTERONISM (CONN'S SYNDROME)

Only two years after the discovery of aldosterone, Conn described the first case of excessive aldosterone secretion arising from an adrenocortical tumour. In almost all the cases which have been reported subsequently, a small yellowish cortical adenoma has been the pathological lesion. In very rare instances an adrenal carcinoma is present and occasionally in children bilateral hyperplasia is found. Incubation of tumour tissue *in vitro* has yielded large amounts of aldosterone and also corticosterone. It is possible that the latter steroid is predominant in some cases of apparent aldosteronism in which the secretion of aldosterone is not very elevated. The tumour is almost always autonomous so that it is not usually susceptible to the normal influences controlling aldosterone secretion, such as sodium depletion. As may be expected, plasma renin concentrations have been found to be low. Conn's syndrome is therefore to be distinguished from excessive aldosterone production which is secondary to conditions such as nephrosis, cirrhosis of the liver with ascites, and cardiac failure, in which renin secretion is often high. Hyperaldosteronism secondary to ischaemic renal disease may lead to particular difficulties in diagnosis and is discussed later.

The diagnosis of primary hyperaldosteronism

While the clinical features may be variable, the diagnosis is usually suspected when benign hypertension is accompanied by a low serum potassium level, usually below 3 m-equiv/litre; the plasma bicarbonate is often raised to 35 m-equiv/litre or more. Conn (1964) has drawn attention to the fact that the serum potassium may be in the low normal range but falls if the daily sodium intake is raised to 200 m-equiv, since aldosterone secretion continues unabated and allows more exchange with potassium at the distal renal tubule. In the normal subject, aldosterone secretion is suppressed by this sodium intake. Since hypokalaemia is an important initial diagnostic finding, it is necessary to exclude other possible causes. Estimation of urinary potassium is first required in order to establish that the hypo-kalaemia is due to excessive renal excretion, and this is likely if urinary levels of greater than 25 m-equiv/daily are found in the presence of a serum potas-sium of, say, 2·5 m-equiv/litre. It should be remembered however, that the most common cause of a low serum potassium level in patients with hyper-tension is the therapeutic use of diuretics, especially of the "thiazide" type. The hypokalaemia in primary hyperaldosteronism is difficult to correct even when the daily potassium intake is raised to 200 m-equiv and this may be a diagnostic pointer. If, however, the sodium intake is reduced to a low level of say 20 m-equiv, and the high potassium is maintained, the serum potassium more easily returns to normal, since there is now less sodium to exchange with potassium in the renal tubule. The serum sodium in Conn's syndrome is often, though not invariably, raised to a moderate degree and Conn has reported a range of 137–160 m-equiv. This finding is of some diagnostic value, since in many instances of secondary hyperaldosteronism associated with oedematous states, the serum sodium tends to be normal or reduced. In about 50% of cases the serum magnesium is below the normal value and it is possible that this factor may contribute to the tetany seen in some patients.

Changes in renal function occur which are of some assistance in diagnosis. Typically there is polyuria due to failure of the concentrating mechanism, which is not responsive to vasopressin, and is probably related to the effects of prolonged hypokalaemia on renal tubular function. The urine has a low osmolality and the pH is neutral or slightly alkaline. These simple bio-chemical changes are not all invariably present, especially in early cases, but they are emphasized because they provide the basis upon which the diagnosis is usually made. The direct demonstration of excessive aldosterone production is difficult for technical reasons. Moreover, estimates of the excretion of aldosterone in the urine, as usually measured by the level of acid-labile conjugate, may not be helpful, since values are sometimes found to be normal and in this respect the finding is analogous to the normal excretion of cortisol metabolites often encountered in Cushing's syndrome. The daily aldosterone secretion rate is almost always raised, but very few laboratories have facilities for carrying out this estimation.

The main diagnostic difficulty in patients with hypertension and a low serum potassium, proved to be of renal origin, is hyperaldosteronism secondary to ischaemic renal disease. In this condition the aldosterone secretion rate is often higher than in Conn's syndrome, where figures greater that 1000 μg daily are unusual. In secondary hyperaldosteronism, plasma renin levels are elevated, whereas in the primary condition they are low. Estimation of plasma renin would therefore be very useful in diagnosis, but, owing to its complexity, the procedure can be carried out in only very few centres. The report of a comparatively simple radioimmunoassay for angiotensin (Boyd *et al.* 1967) is therefore of interest in this respect. The diagnosis of secondary hyperaldosteronism is usually made by the demonstration of underlying renal disease by such radiological aids as pyelography and angiography. In malignant hypertension there is often a tendency to hypokalaemic alkalosis, even in the absence of diuretic therapy, and in many cases a high aldosterone secretion has been found. The elevated renin levels, however, indicate that the aldosteronism is of renal origin. In Conn's syndrome the hypertension is almost always of the benign variety, and in the uncommon event of malignant hypertension being present it is likely that ischaemic renal disease has supervened.

Conn has of recent years attempted to incriminate as many as 20% of cases of benign hypertension, with or without hypokalaemia, as examples of his syndrome, using as ancillary evidence the fact that a large number of adrenal adenomata are found accidentally in hypertensive patients who have a postmortem examination. Subsequent work has failed to show, however, either a raised aldosterone secretion in ordinary benign hypertension, or an abnormally high concentration of aldosterone in these adenomata, the great majority of which are functionless.

Detailed information on primary and secondary hyperaldosteronism is to be found in *Aldosterone*, a symposium edited by Baulieu & Robel (1964).

REFERENCES

BAULIEU, E. E. and ROBEL, R. (1964). *Aldosterone*. Blackwell, Oxford.
BOYD, G. W., LANDON, J., and PEART, W. S. (1967). *Lancet* **2**, 1002–1005.
CONN, J. W., KNOPF, R. E., and NESBIT, R. M. (1964) *Aldosterone*, pp. 327–352. Ed. by E. E. BAULIEU and P. ROBEL. Blackwell, Oxford.
COPE, C. L. and PEARSON, J. (1965). *J. clin. Path.* **18**, 82–87.
FARMER, T. A., HILL, S. R., PITTMAN, J. A., and HEROD, J. W. (1961). *J. clin. Endocr.* **21**, 433–455.
GRABER, A. L., NEY, R. L., NICHOLSON, W. E., ISLAND, D. P., and LIDDLE, G. W. (1965). *J. clin. Endocr.* **25**, 11–16.
HUTTER, A. M. and KAYHOE, D. E. (1966). *Amer. J. Med.* **41**, 572–580, 581–592.
JENKINS, J. S. (1968). In *Memoirs of the Society for Endocrinology*, No. 17. pp. 205–212. Ed. by V. H. T. JAMES and J. LANDON, Cambridge University Press.
LIDDLE, G. W., GIVENS, J. R., and NICHOLSON, W. E. (1965). *Cancer Res.* **25**, 1057–1061.
LIVANOU, T., FERRIMAN, D., and JAMES, V. H. T. (1967). *Lancet*, **2**, 856–859.

F

MIGEON, C. J., GREEN, O. C., and ECKERT, J. P. (1963). *Metabolism*, 718–739.
NELSON, D. H., MEAKIN, J. W., and THORN, G. W. (1960). *Annal. Int. Med.* **52**, 560–569.
ROSS, E. J., MARSHALL-JONES, P., and FRIEDMAN, M. (1966). *Quart. J. Med.* **35**, 149–192.
WILLIAMS, R., KELLIE, A. E., WADE, A. P., WILLIAMS, E. D., and CHALMERS, T. M. (1961). *Quart J. Med.* **30**, 269–284.

PART III

Laboratory Procedures

10

THE ESTIMATION OF ADRENAL STEROIDS IN URINE

In the investigation of disorders of the adrenal cortex most laboratories rely to a large extent on the estimation of urinary steroids because generally the techniques are less exacting than those required for the determination of steroids in blood and they also provide a measure of adrenocortical function over the whole period of 24 hours. A major disadvantage of urinary estimations, however, is that they represent mainly the metabolites of adrenal secretion, so that abnormalities of metabolism or renal function may determine the values irrespective of the actual levels of hormone in the circulation. Most routine urinary methods estimate whole groups of steroid compounds; the determination of individual steroids requires separation by chromatographic techniques..

URINARY 17-OXOSTEROIDS (17-KETOSTEROIDS)

The 17-oxosteroids all have a ketone group at C-17 and the term generally denotes compounds with 19 carbon atoms. They comprise a mixture of metabolites derived not only from secretion of the adrenal cortex but also the testes, and, to a small extent, the ovaries. In the male, two-thirds arise from the adrenal cortex. Although the 17-oxosteroids were the first steroids to be determined by a reasonably satisfactory chemical technique, they do not provide a specific measure of cortisol secretion and the main use of the estimation is now in the investigation of virilizing disorders, but even here it is of limited value. The 17-oxosteroids can be divided into those which have an oxygen atom at the 11-position and those in which it is absent. The latter, 11-deoxy-17 oxosteroids, form the larger proportion and comprise mainly dehydroepiandrosterone, androsterone and aetiocholanolone, of which the last two are metabolites of both adrenal and testicular secretion. The 11-oxy-17-oxosteroids are entirely adrenal in origin and are formed partly from the oxidation of the side chain of cortisol, of which only about 5% is broken down in this manner, and partly from the metabolism of the adrenal C_{19} compound 11β-hydroxyandrostenedione. The 17-oxosteroids are present in urine conjugated either with glucuronic acid or with sulphate and the initial step in their estimation is the hydrolysis of the conjugates to the free steroids.

Principle of method for the estimation of 17-oxosteroids

The technique usually employed in Great Britain is similar to that recommended by a Medical Research Council Committee on Clinical Endocrinology (1963). Hydrolysis is carried out by boiling a 10 ml. volume of a 24-hour urine sample with strong hydrochloric acid. The steroids are then extracted with an organic solvent such as ethylene dichloride or chloroform, and after washing the extract with alkali, the 17-oxosteroids are estimated colorimetrically by the Zimmerman reaction. In this reaction *m*-dinitrobenzene in alcoholic potassium hydroxide reacts with 17-oxosteroids to give a deep purple colour having a maximum light absorption at 520 mμ. A solution of dehydroepiandrosterone is generally used as the standard although individual 17-oxosteroids give different colour intensities with the Zimmerman reagent. In practice, however, the overall error introduced into the estimation of the mixture of 17-oxosteroids is not very great. Various non-steroid substances in urine extracts also yield colours with the Zimmerman reagent, so that a correction factor must be applied. The best results are obtained by the use of a spectrophotometer and the correction of Allen (1950), for which readings equidistant from the 520 mμ peak are also taken, usually at 440 mμ and 600 mμ. The corrected extinction E is then given by the formula,

$$E = E_{520} - \left[\frac{E_{440} + E_{600}}{2} \right]$$

Even after application of the correction factor, however, it is likely that variable amounts of non-specific chromogen contribute to the result. The method is relatively simple to perform and is also the basis of the 17-oxogenic steroid estimation described below.

Normal values. In males aged 20–40 the normal range is 10–25 mg daily, and in females 5–17 mg, the difference being due to the testicular component. There is a gradual decline with increasing age. In children, Prout & Snaith (1958) give the following mean figures: 0–1 yr, 0·3 ± 0·1; 1–5 yr, 0·8 ± 0·5; 6–10 yr, 1·4 ± 0·7; 11–17 yr, 5·0 ± 2·1. The precision of the method below 2 mg daily is, however, not very high, owing to the presence of non-specific chromogen.

Low values are encountered in hypoadrenal states, either primary or secondary to hypopituitarism, but it should be emphasized that low values are also found in many wasting conditions such as anorexia nervosa, and in hypothyroidism.

Raised values are seen after the administration of ACTH to normal individuals, but the change is very variable and does not provide a reliable guide to the adrenal cortical response to ACTH. In Cushing's syndrome due to adrenal hyperplasia the 17-oxosteroids may be raised but in one large series a range of 2·6–50 mg daily was observed, with a mean of 15·2 mg. In four cases of adenoma, the range was 5–26 mg. In adrenal carcinoma, however, values are often very high, and figures ranging from 50 to 1000 mg daily have been reported. In congenital adrenal hyperplasia, raised levels of urinary 17-oxosteroids are often used as an important diagnostic aid, but in mild cases of C-21 hydroxylase deficiency, or cases of the 11β-hydroxylase defect, values may not be greatly increased. In some cases of adult virilism and in many patients with hirsutism only, the total 17-oxosteroids are within the normal range, since small but biologically significant increases in potent androgens may not be accompanied by a measurable increase in urinary metabolites.

Individual 17-oxosteroids. Fractionation of the 17-oxosteroids into 11-deoxy and 11-oxy groups can be carried out by fairly simple column chromatography, although for satisfactory results the preliminary hydrolysis of conjugates must be by more gentle means than the vigorous hot acid technique used for the total 17-oxosteroid estimation. Separation into individual steroids can also be performed by paper, thin layer, or, more recently, gas-liquid chromatography (Chapter 13). The normal range for individual 17-oxosteroids in urine as reported by Jailer *et al.* (1959) is shown in Table 10.1. Recent methods using gas-liquid chromatography give somewhat lower values for the 11-deoxy steroids.

TABLE 10.1. Normal range of individual urinary 17-oxosteroids.

		Mg per 24 hours	
		Male	Female
11-deoxy-17-oxo-steroids	Androsterone	1·8–5·7 (3·6)	0·9–3·9 (2·0)
	Aetiocholanolone	0·8–6·5 (3·1)	0·6–3·8 (2·0)
	Dehydroepiandrosterone	0·1–8·8 (2·7)	
11-oxy-17-oxo-steroids	11β-hydroxy-androsterone	0·6–3·5 (1·5)	0·5–1·7 (1·0)
	11β-hydroxy-aetiocholanolone	0·2–0·9 (0·5)	0·0–1·5 (0·5)
	11-oxo-aetiocholanolone	0·2–1·2 (0·6)	0·0–1·0 (0·5)

Mean figures are given in parentheses.

ESTIMATION OF URINARY METABOLITES OF CORTISOL

Of much greater general importance than the 17-oxosteroids in the study of adrenal cortical function is the estimation of the C_{21} compounds in urine, particularly those metabolites derived from cortisol. The methods described below are dependent upon reactions involving the side chain attached to carbon 17, and therefore measure groups of steroids which possess the particular side chain. There are two techniques in general use at the present time.

(1) Total 17-oxogenic steroids (total 17-hydroxycorticosteroids)

The most widely used method for the estimation of C_{21} compounds in Great Britain is the 17-oxogenic steroid technique, devised by Norymberski in 1952, and the subsequent improvements have been reviewed by him in detail (Norymberski, 1961). The basis of the method is oxidation of the side chain at C-17 to form a 17-oxosteroid which is then estimated by the Zimmerman reaction. In the original technique the pre-existing 17-oxosteroids were estimated separately and subtracted from the total in order to obtain the value for the 17-oxosteroids newly formed from oxidation of the C_{21} compounds. In the later, improved version, due to Appleby, Gibson, Norymberski, and Stubbs, the pre-existing 17-oxosteroids are first removed and the C_{21} compounds can then be estimated directly. This estimation was originally designated "total 17-hydroxycorticosteroids" but this term is better reserved for an entirely different procedure described below, and, following the recommendations of the Medical Research Council Committee (1963), is called "total 17-oxogenic steroids".

Principle of method for estimation of total 17-oxogenic steroids

4 ml. from a 24-hour sample of urine is treated with potassium borohydride overnight to reduce the 17-oxosteroids already present. After acidification the side chain of the C_{21} compound is oxidized with sodium bismuthate and the Zimmerman reaction is them applied as already described. The types of compounds and the side chains at C-17 which take part in this procedure are shown in Fig. 10.1. Metabolites of cortisol, mainly tetrahydrocortisol, tetrahydrocortisone, the cortols, and the cortolones, ordinarily constitute the majority of the 17-oxogenic steroids but it will be noted that pregnanetriol can also take part in the reaction. In congenital adrenal hyperplasia, where cortisol synthesis is defective, an elevated excretion of 17-oxogenic steroids is found but is mainly due to the presence of pregnanetriol. No special measures are required for the hydrolysis of conjugates since the oxidation by bismuthate breaks the glucuronide linkage and allows extraction by organic solvents. The main disadvantage of the method is variation in the oxidation of the side chain encountered with different batches

of bismuthate, together with further uncontrolled oxidation which may change the chromogenic intensity in the Zimmerman reaction. For these reasons, Few (1961) substituted sodium periodate for bismuthate as the oxidizing agent and in the experience of the writer Few's technique is much

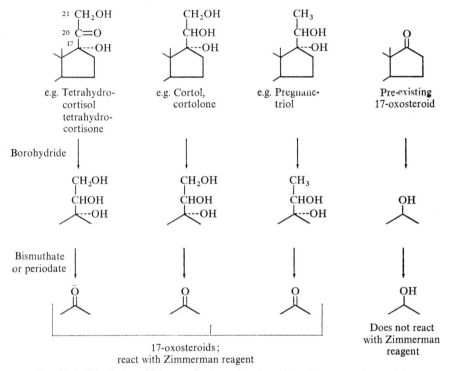

FIG. 10.1. Side chains which react in the estimation of total 17-oxogenic steroids.

preferable. A shorter and simpler version of the periodate method has been described by Metcalf (1963) which has been reported to be satisfactory for routine purposes by several groups of workers. The presence of glucose in the urine leads to false values being obtained by all of the techniques. Norymberski suggests increasing the amount of bismuthate or removing the glucose by yeast fermentation. Few recommends diluting the urine to a total volume of 4 litres before the estimation, or the conjugates can be extracted from the urine by an ether-alcohol mixture before oxidation. A useful extension of the 17-oxogenic steroid procedure, based upon a technique of Morris for the estimation of pregnanetriol, has been described by Hill (1960) for the diagnosis of C-21 hydroxylase deficiency and is referred to in Chapter 7. In this technique the ordinary procedure is first performed and the resulting

17-oxosteroids are separated by column chromatography into 11-oxy and 11-deoxy fractions. The product from pregnanetriol appears in the 11-deoxy fraction so that in congenital adrenal hyperplasia the ratio of 11-deoxy to 11-oxy is usually greater than 1·0 whereas in the normal child the ratio is less than 0·5. It has been found, however, that very young infants with the defect tend to have a lower ratio than the older children owing to a higher excretion of 11-oxopregnanetriol in early life.

Normal values for 17-oxogenic steroids range in adult males from 5 to 21 mg per 24 hours with a mean of 11 mg, and in females from 4 to 16 mg with a mean of 9 mg. There is some decline with increasing age but this is less than is the case with 17-oxosteroids. In children under the age of 10 the range is 2·3–3·8 mg per 24 hours.

Low values are found in hypoadrenal states whether arising from Addison's disease or hypopituitarism. There is, however, an appreciable number of patients with hypoadrenalism who have levels in the normal range, and of much greater diagnostic value is the response to the administration of ACTH or metyrapone described in Chapter 12. Values lower than normal are also found in hypothyroidism and in hepatic cirrhosis owing to the decreased metabolism of cortisol in these conditions.

Raised values are found in Cushing's syndrome but in some cases the levels fall within the normal range and there is considerable day to day variation. In one large series of patients with Cushing's syndrome only 35% had 17-oxogenic steroids which were consistently above the normal range. Moreover, moderately raised levels are sometimes found in simple obesity. At high levels of adrenal activity the 17-oxogenic steroids represent very approximately 50% of the cortisol secreted, but Cope and Pearson (1965) have emphasized the relatively poor correlation between the 17-oxogenic steroid excretion and cortisol secretion rate in the near normal range. Nevertheless, the estimation of the urinary 17-oxogenic steroids provides a useful measure of adrenocortical function within the capability of a routine laboratory when it is used in conjunction with stimulation or suppression tests.

(2) Estimation of urinary 17-hydroxycorticosteroids (Porter-Silber Method)

In the U.S.A. a widely used technique for the estimation of urinary metabolites of cortisol is dependent upon the reaction of the 17,21-dihydroxy-20-oxo group with phenylhydrazine in strong sulphuric acid to give a yellow colour with a maximum light absorption at 410 mμ. This constitutes the Porter-Silber reaction and has also had a wide application in the estimation of cortisol in blood. The urinary method most employed is based upon that of Glenn & Nelson (1953) and requires, as the first step, enzymatic hydrolysis of the conjugates, followed by extraction with chloroform, or, better, methylene chloride, an alkaline wash, and finally extraction into the phenylhydrazine reagent. Only steroids with the 17,21-dihydroxy-20-oxo side chain are estimated, so that the cortols, cortolones, and pregnanetriol are not included. Normal values are therefore lower than those given for the 17-oxogenic steroids and range from 4·1 to 14 mg in males and 2·8 to 12 mg in females. The method is useful as a routine procedure, but suffers from the disadvantage that preliminary enzyme hydrolysis is necessary. A comparison of the 17-oxogenic and Porter-Silber methods is given by Golub *et al.* (1958).

Estimation of urinary pregnanetriol

The main adrenal source of pregnanetriol is 17α-hydroxyprogesterone (see Chapter 4) and its estimation in urine is an important procedure in the diagnosis of C-21 hydroxylase deficiency. The method recommended by a Medical Research Council Committee (1963) is essentially due to Bongiovanni and Eberlein. Pregnanetriol is present in urine as a glucuronide and hydrolysis must first be carried out by the use of the enzyme β-glucuronidase. Extraction of the free steroid is then performed, preferably with benzene, and after an alkaline wash the extract is chromatographed on a column of alumina, using 10% ethanol in benzene to elute the pregnanetriol from the column. The final estimation is carried out colorimetrically by adding concentrated sulphuric acid and reading the peak absorption at 435 mμ. Readings are also taken at 400 and 470 mμ in order to employ the Allen correction, in a manner similar to that described in the 17-oxosteroid estimation. Alternatively, instead of using the sulphuric acid chromogen after elution from the column, the pregnanetriol can be oxidized by sodium periodate to aeticholanolone which is estimated by the Zimmerman reaction. The latter technique is to be preferred since it is rather more specific and more sensitive (Harkness & Love, 1966).

Normal values. In children values are less than 0·15 mg daily. In adult males the range is 0·4–2·4 mg daily. In females there is some variation with the menstrual cycle, being maximal in the luteal phase, when the range is 0·5 to 2·8 mg daily.

Raised values are found in congenital adrenal hyperplasia most often due to deficiency of C-21 hydroxylase although in a few cases values within the normal range have been reported. Raised levels are also found in adrenal carcinoma where the precursor is probably 17-hydroxypregnenolone.

REFERENCES

ALLEN, W. M. (1950). *J. clin. Endocr.* **10**, 71–83.
COPE, C. L. and PEARSON, J. (1965). *J. clin. Path.* **18**, 82–87.
FEW, J. D. (1961). *J. Endocr.* **22**, 31–46.
GLENN, E. M. and NELSON, D. H. (1953). *J. clin. Endocr.* **13**, 911–921.
GOLUB, O. J., SOBEL, C., and HENRY, R. J. (1958). *J. clin. Endocr.* **18**, 522–530.

HARKNESS, P. A. and LOVE, D. N. (1966). *Acta. Endocr. (Kbh.)* **51**, 526–534.

HILL, E. (19b). *Acta. Endocr. (Kbh.)* **33**, 230–250.

JAILER J. W., 'ANDE WIELE, R., CHRISTY, N. P., and LIEBERMAN, S. (1959). *J. clin. Invest.* **38**, 357–365.

MEDICAL RESEAR 'H COUNCIL COMMITTEE ON CLINICAL ENDOCRINOLOGY, (1963). *Lancet*, **1**, 1415–1419.

METCALF, M. (1963, *J. Endocr.* **26**, 415–423.

NORYMBERSKI, J. K. (1961). In *The Adrenal Cortex*, pp. 88–109. Ed. by G. K. McGOWAN and M. SANDLER. Pitman, London.

PROUT, M. and SNAITH A. H. (1958). *Arch. Dis. Childh.* **33**, 310–304.

11

THE ESTIMATION OF CORTISOL
IN BLOOD

Of recent years techniques for the estimation of cortisol levels in blood have been devised which are within the capabilities of ordinary laboratories. These methods complement and do not replace urinary estimations since the values obtained only refer to the single moment of time at which the sample was taken. In order to obtain information of the state of adrenal function over the whole period of 24 hours, estimation of urinary 17-oxo-genic steroids is necessary or, better, the cortisol secretion rate if the facilities are available. Nevertheless, blood estimations are being widely used, owing to the speed with which they can be performed and because they are less affected by changes in steroid metabolism. The normal diurnal variation makes it necessary to know the time of obtaining the blood sample and ordinarily this is taken between 8 and 10 a.m. when the levels are near maximal. Before describing these relatively simple techniques it is important to be aware, firstly, that they are not completely specific for cortisol, and secondly, that their apparent simplicity is deceptive; extreme attention to details especially in the purification of reagents and cleanliness of laboratory equipment is required in order to obtain satisfactory results.

The estimation of plasma 11-hydroxycorticosteroids by fluorimetric methods

The method most in favour in Great Britain at the present time is a fluorimetric technique devised by De Moor et al. (1960) and further simplified by Mattingly (1962).

Principle of method (Mattingly)
2 ml of heparinized plasma is extracted with methylene chloride which has been specially redistilled for the purpose. The methylene chloride phase is separated and allowed to react for thirteen minutes with a fluorescence-inducing reagent which consists of 70% concentrated sulphuric acid and 30% ethanol. The optimum wavelength of the light exciting the fluorescence

87

is 470 mμ and the emitting wavelength is 530 mμ. The fluorescence is read in a suitable fluorimeter, corrected for a reagent blank, and compared with a cortisol standard. Under these conditions both cortisol and corticosterone are estimated, hence the term plasma "11-hydroxycorticosteroids" instead of "cortisol" is more correct, although ordinarily corticosterone is present in only about one-tenth the concentration of cortisol. Most corticosteroids such as tetrahydrocortisol, 11-deoxycortisol, cortisone, and synthetic derivatives, do not fluoresce significantly.

Non-specific fluorescence

While such an extraction procedure is simple, its disadvantage lies in the presence of non-steroid substances in the plasma which contribute to the fluorescence and lessen the specificity of the method. This non-specific fluorescence varies greatly in different plasma samples, even in patients not receiving treatment with drugs. It can be estimated in plasma obtained from cases of severe Addison's disease or after bilateral adrenalectomy. Mattingly (1962) gives a range of 0–4·2 μg per 100 ml. in patients with signs of acute adrenal insufficiency but in the experience of many workers, the Mattingly modification rarely gives values below 5 μg even in adrenalectomized subjects. This non-specific fluorescence is of less importance in the middle ranges of normal and for the assessment of the response to ACTH, but basal values found to be within the lower normal range, 6–10 μg per 100 ml., cannot exclude adrenal insufficiency. Attempts have therefore been made to improve the specificity. Firstly, it is necessary to use the optimum exciting wavelength of 470 mμ which cannot be achieved with the fluorimeter originally recommended in the Mattingly procedure. A spectrofluorimeter gives good results but this instrument is very expensive and there are now direct-reading instruments using a zinc lamp source, which with suitable filters gives the correct wavelength. Even with better instrumentation, however, James *et al.* (1967) in a careful comparison of the Mattingly procedure with a double isotope technique specific for cortisol, showed that as much as 25–55% of fluorescence was due to substances other than cortisol. Corticosterone only contributes a small proportion of this. James *et al.* suggest increasing the strength of the acid in the fluorescence reagent to 80% and shortening the time of reaction to 5 minutes. This reduces the non-specific element to much lower levels but requires a semi-automated procedure. Spencer-Peet *et al.* (1964) observed that the non-steroid fluorescence increased progressively with time, whereas the fluorescence of cortisol remained constant over the range of 8–16 minutes. They concluded that the intercept produced by extrapolation of the fluorescence slope to zero time gave the proportion of the observed fluorescence due to the steroids. In some plasmas the slope is steep and in others there is comparatively little change. An example of the fluorescence obtained using a sample of plasma from an adrenalectomized subject with and without the addition of cortisol is shown in Fig. 11.1. The intercept can be obtained by a simple mathematical calculation if readings are made at 8 minutes and 16 minutes;

Corrected fluorescence=(fluorescence at 8 min × 2)−fluorescence at 16 min.

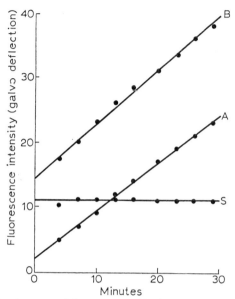

FIG. 11.1. Development of fluorescence with time.
A = sample of plasma from adrenalectomized subject.
B = A + cortisol (10 μg per 100 ml.)
S = Standard cortisol (10 μg per 100 ml.)

This correction depends upon the fact that the increase in non-specific fluorescence with time is invariably linear and this may not necessarily be so in every case. Nevertheless, in the experience of the writer this is the most useful of the simpler procedures designed to improve specificity in the low range since zero levels are frequently obtained in adrenalectomized patients. More work on improvements in fluorimetric techniques is, however, required. *Normal values* between 9 and 10 a.m. are given by Mattingly as 6·5 μg to 26·3 μg per 100 ml. (mean 14·2). In the writer's laboratory employing the Spencer-Peet modification the normal range is 5·5 to 23·2 μg at 10 a.m. (mean 14·0 ± 4·1 S.D.). Values of 1·1 to 8·5 μg are found at midnight.

Estimation of plasma 17-hydroxycorticosteroids (Porter-Silber colorimetric method)

In the U.S.A. the method most often used for the estimation of cortisol in blood is based upon the Porter-Silber reaction of acid phenylhydrazine with the 17,21 dihydroxy-20-oxo side chain and is essentially similar to the method employing this reaction for the estimation of urinary corticosteroids. In the first successful assay of cortisol in blood to be devised, due to Nelson and Samuels, preliminary column chromatography was used, but later, Silber and Porter dispensed with the need for chromatography and their method was further improved by Peterson *et al.* (1957). The modification of Peterson is the most widely used technique.

Principle of method. A minimum of 5 ml. of heparinized plasma is required. This is extracted with methylene chloride followed by a wash with sodium hydroxide and the methylene chloride phase is further extracted with 0·5 ml.

TABLE 11.1. Comparison of the fluorimetric and colorimetric methods for the estimation of plasma cortisol

	Fluorimetric (Plasma 11-hydroxycorticosteroids)	Colorimetric (Plasma 17-hydroxycorticosteroids)
Sample volume	1–2 ml.	At least 5 ml.
Working time	Six samples in 1–2 hours	Six samples in 3–4 hours and overnight colour development
Specificity	Less interference with drugs but variable amount of non-specific fluorescence in plasma.	Drugs interfere, especially paraldehyde. Probably less non-specific colour from plasma only.
Steroids estimated	Cortisol and corticosterone. *Not* cortisone, 11-deoxycortisol or synthetic glucocorticoids.	17-21 dihydroxy-20-oxo steroids e.g. cortisol, cortisone, 11-deoxy-cortisol. *Not* corticosterone.

of phenylhydrazine in 64% sulphuric acid-ethanol. The upper methylene chlorid elayer is discarded and the phenylhydrazine reagent is allowed to stand overnight for the colour to develop. Readings are taken at 410 mμ in a spectrophotometer capable of taking micro-cells, and corrections are applied for sample and reagent blanks. Cortisol standards are taken through the whole procedure.

The normal range between 9 and 10 a.m. is 6 to 25 μg per 100 ml. (mean 13·0 ± 6·0 S.D.) (Jenkins, 1961). A comparison of the commonly used fluorimetric and colorimetric techniques is shown in Table 11.1. The fluorimetric method is more sensitive, is about three times as rapid to perform, and can be used to estimate cortisol in the presence of other 17-hydroxycorticosteroids. On the other hand, the colorimetric method can be used for the estimation of plasma 11-deoxycortisol in the metyrapone test, and is probably rather more specific in the low range, if steps are taken to discontinue drug therapy.

The estimation of plasma cortisol by competitive protein-binding radio-assay

A completely different approach to the estimation of steroids in blood has been suggested by Murphy (1967). In this ingenious technique the ability of a free steroid to displace protein-bound steroid from its attachment to corticosteroid-binding globulin is used. The proteins in the plasma to be assayed are first removed by such measures as ethanol precipitation. The cortisol present is then allowed to mix with a solution of corticosteroid-binding globulin which is just saturated with tritium-labelled cortisol. The steroid in the sample displaces a part of the isotopically-labelled steroid and the percentage of tracer bound to corticosteroid-binding globulin falls proportionately. The protein-bound and free steroid are separated by a simple technique which involves absorption of the free steroid on a silicate suspension. A standard curve is obtained by plotting percentage binding against known concentration of cortisol, and the amount in the sample is then read off the curve. Extreme sensitivity together with specificity are claimed for this technique. Normal values of 6 to 18 μg per 100 ml. are given by Murphy using volumes of 0·1 ml. of plasma. The precision is claimed to be good but more experience is required from other workers. Other methods for the estimation of plasma cortisol involving its isolation by paper chromatography are specific but too time-consuming for routine use.

The value of plasma cortisol estimations

The estimation of cortisol by the fluorimetric technique outlined above can provide a very rapid assessment of adrenal cortical function, avoiding the need for 24-hour urine collections, so that it can be used even on an out-patient basis. It is of most value when used in conjunction with procedures for suppression or stimulation of the adrenal cortex. A comparison between morning and night levels is often informative. Plasma cortisol estimations also provide the measure of response to relatively short-lived stimuli such as vasopressin, pyrogen, and insulin hypoglycaemia described in Chapter 12. On the other hand when the metyrapone test is used, the fluorimetric method is not possible and the estimation of urinary steroids is then required or estimation of plasma 17-hydroxycorticosteroids by the Porter-Silber technique.

Low values. The lack of specificity is most apparent for levels below 10 μg per 100 ml. and for this reason single estimations are of little value in the diagnosis of adrenal insufficiency. When this is suspected an ACTH test should always be carried out.

High values may be obtained spuriously if the technique is not competently performed. Truly high figures are obtained in pregnancy since in this condition there is a progressive increase in corticosteroid-binding-globulin and the total bound and free steroid is estimated. In the last trimester figures of

G

25 to 50 µg per 100 ml. may be obtained. Similarly, high values are found in non-pregnant women who are receiving oral contraceptives owing to the oestrogen which they contain.

In Cushing's syndrome, higher than normal figures may be obtained but values are sometimes found to be normal even at 8 to 10 a.m. when levels are at their maximum. A sample taken at midnight is useful since in this condition the normal diurnal variation is often lost. The effect of dexamethasone on the plasma cortisol level is also a useful diagnostic procedure.

In the interpretation of raised levels of plasma cortisol, the effect of stresses such as anxiety, and traumatic conditions, including surgery, should be remembered.

In infancy immediately after delivery the level of cortisol in the blood is high due to the transplacental passage of some maternal cortisol, after which it rapidly falls so that values of 2 to 4 µg are found by the third day. Thereafter the levels gradually rise and reach the normal concentration seen in the adult at about 10 days.

REFERENCES

DE MOOR, P., STEENO, O., RASKIN, M., and HENDRIKX, A. (1960). *Acta. Endocr.* (Kbh.) **33**, 297–307.
JAMES, V. H. T., TOWNSEND, J., and FRASER, R. (1967). *J. Endocr.* **37**, xxviii.
JENKINS, J. S. (1961). In *The Adrenal Cortex.* pp. 33–46. Ed. by G. K. MCGOWAN and M. SANDLER. Pitman, London.
MATTINGLY, D. (1962). *J. clin. Path.* **15**, 374–379.
MURPHY, B. E. P. (1967). *J. clin. Endocr.* **27**, 973–990.
PETERSON, R. E., KARRER, A., and GUERRA, S. L. (1957). *Analyt. Chem.* **29**, 144–149.
SPENCER-PEET, J., DALY, J. R., and SMITH, V. (1965). *J. Endocr.* **31**, 235–244.

12

THE ASSESSMENT OF HYPOTHALAMIC-PITUITARY-ADRENAL FUNCTION

The estimation of corticosteroids in blood or urine provides the basis for a series of procedures designed to test the function of the adrenal cortex and its connections with the anterior pituitary and hypothalamus. (Fig. 12.1).

THE CORTICOTROPHIN (ACTH) TEST

In order to test the ability of the adrenal cortex to secrete cortisol, the response to the administration of ACTH in maximally stimulating doses is determined. Preparations of ACTH commonly available include:

Porcine ACTH—(soluble) for intravenous or intramuscular use.

Porcine ACTH-zinc hydroxide suspension for intramuscular use only.

Porcine ACTH in gelatine for intramuscular use only.

Synthetic ACTH (Synacthen) soluble for intravenous or intramuscular use.

Synthetic ACTH-zinc syspension for intramuscular use.

The synthetic preparations contain only the N-terminal 24 amino acids of ACTH and have the advantage of less risk of provoking allergic reactions, especially in subjects with Addison's disease; 1 mg is approximately equal to 100 international units of porcine ACTH. A single intramuscular injection of soluble ACTH results in a maximal adrenal response at about 1 hour, but the effective duration is only about 2–3 hours. Corticotrophin gel or zinc has a maximal effect at about 5–6 hours and persists for up to 24 hours. The ACTH test is most often carried out by determining the increase in urinary 17-oxogenic steroids or Porter-Silber 17-hydroxycorticosteroids. Two control 24-hour urine specimens are obtained following which 40 units of ACTH is administered intramuscularly as the zinc or gel preparation twice daily for two to three days, and the level of urinary corticosteroids during these days is determined. The normal figure on the first day is at least 30 mg of 17-oxogenic steroids, and higher levels are found on succeeding days of stimulation. Alternatively, 25 units of soluble ACTH can be given as an intravenous infusion in normal saline over a period of 6 to 8 hours on two successive days. The latter route of administration should be used to repeat the test if an equivocal result has raised any doubt concerning the adsorption of ACTH by the intramuscular route. A more rapid procedure is provided by the use of the plasma corticosteroid level as the indicator of response.

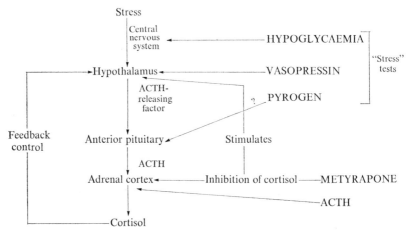

Fig. 12.1. Tests of hypothalamic-pituitary-adrenal function and their probable sites of action.

A sample of blood is taken in the morning and again 5 hours after an intramuscular injection of ACTH-gel or zinc, or at the end of a six hour intravenous infusion of soluble ACTH. In cases of suspected hypopituitarism the procedure should be repeated on a further one or two days. In normal individuals, using the fluorimetric method of estimation of plasma corticosteroids, the increase 5 hours after 40 units of ACTH-zinc ranged from 10 to 50·8μg with a mean figure of 26·0 ± 4·7 S.D.

Rapid ACTH test. The plasma corticosteroid response to soluble ACTH can provide the basis of a very rapid screening procedure for suspected adrenal insufficiency, which is suitable for outpatients. A control blood sample is taken following which 25 units (1·25 mg) of synthetic ACTH is given intramuscularly, and 30 minutes later a second blood sample is obtained. A mean increase of 16·7 ± 7·7 μg per 100 ml. (range 7·5 to 27·5) was obtained by Wood *et al.* (1965). A normal increase excludes adrenal insufficiency, but if a subnormal response is obtained one of the prolonged ACTH tests outlined above must be carried out to confirm Addison's disease and to detect the sluggish response found in some cases of hypopituitarism.

Interpretation and Use

The ACTH test finds most use in the diagnosis of adrenocortical insufficiency. In Addison's disease control steroid levels may be very low or sometimes within the normal range, but the absence of response to a properly conducted ACTH test is diagnostic of the condition. In severe hypopituitarism control levels are below normal but a sluggish response is usually obtained if ACTH is prolonged for several days. In milder cases of anterior pituitary deficiency the response may be within the normal range but it is necessary to perform the test before proceeding to the measures for testing pituitary-adrenal function described below.

ACTH tests have also been used in the diagnosis of hyperfunction of the adrenal cortex. Thus, in Cushing's syndrome due to adrenal hyperplasia, a greater than normal response may be obtained, but the range of normal is so wide that there are difficulties in interpretation of the hyper-active response. An ACTH test may be useful in distinguishing those cases of Cushing's syndrome due to hyperplasia from those which are outside pituitary control due to adrenal carcinoma or due to extra-adrenal malignant disease. These cases, usually, but not invariably, fail to respond to exogenous ACTH.

In virilizing conditions of adrenal origin not due to malignancy, the administration of ACTH is followed by an increase in adrenal androgens. This can be demonstrated by finding disproportionate increase in urinary 17-oxosteroids, or, much better, an increase in the secretion of individual androgens.

It must be emphasized that an ACTH test only gives an indication of the ability of the adrenal cortex to respond to stimulation. In the investigation of anterior pituitary deficiency, other tests are required to determine the ability of the pituitary to release endogenous ACTH.

THE ASSESSMENT OF PITUITARY-ADRENAL FUNCTION

While a direct measurement of the level of ACTH in the blood would be desirable, technical difficulties preclude this estimation in most laboratories, so that changes in the blood or urinary corticosteroids are used as the indicator of endogenous ACTH release. This approach is valid only if the adrenal cortex is first shown to be fully responsive to ACTH by means of an ACTH test, and this should always be carried out in suspected hypopituitarism.

The metyrapone test

The action of metyrapone ("Metopirone", SU 4885) as an inhibitor of 11β-hydroxylation is described in Chapter 7. The inhibition of cortisol synthesis induced by the drug is normally followed by a release of ACTH which stimulates the adrenal to produce large amounts of 11-deoxycortisol. This compound and its metabolite tetrahydro-11-deoxycortisol in urine is estimated as a 17-oxogenic steroid or 17-hydroxycorticosteroid. The test is thus a measure of the integrity of the feedback control of ACTH release. Two control 24-hour urine specimens are obtained following which metyrapone is administered orally in the dosage of 750 mg 4-hourly for six doses. Urine is collected on the day of metyrapone and on the following day for the estimation of 17-oxogenic steroids or Porter-Silber 17-hydroxycorticosteroids. The maximum increase usually occurs on the day after metyrapone. There has been much discussion as to the normal response. Liddle *et al.* (1959) defined the response as being at least double the control figure. In our laboratory the increase in urinary 17-oxogenic steroids above the control on the day after metyrapone ranged from 9 to 51·0 mg per 24 hours. Porter-Silber 17-hydroxycorticosteroids give a smaller increase. The oral metyrapone test can be carried out more rapidly if the estimation of plasma 17-hydroxycorticosteroids is used. A blood sample is taken in the morning and again at the same time on the following day after the administration of metyrapone. The estimation of plasma corticosteroids must be carried out by a colorimetric technique based upon the Porter-Silber reaction; fluorescent methods are not suitable since they do not measure 11-deoxycortisol. Jenkins & Elkington (1964) found that the normal increase in plasma 17-hydroxycorticosteroids after oral metyrapone, 750 mg 4-hourly for six doses, ranged from 8·0 to 38·0 µg per 100 ml. (mean 19·9 ± 7·2 S.D.). Metyrapone has also been administered as an intravenous infusion by some workers but the oral route provides a more sustained effect and is more convenient.

Side effects. Some patients experience dizziness and especially dyspepsia, which can be minimized by giving the drug with meals or with milk. There is a theoretical risk of producing acute adrenal failure in patients with severe hypopituitarism. Reported instances are rare, but any sudden fall in blood pressure should be the indication for the administration of cortisol intravenously.

Interpretation

The metyrapone test is a measure of the feedback control, the site of which lies at the level of the hypothalamus. Thus, not only lesions of the pituitary itself but also some disorders of the central nervous system may give a subnormal response. The response to the metyrapone test is more often abnormal than in any of the procedures described as "stress tests". Certain drugs which are active on the central nervous system, such as phenytoin and chlorpromazine, have been reported as preventing the response to metyrapone. Some workers have found that myxoedema is associated with a subnormal response. There is undoubtedly a wide individual variation in the inhibiting capacity of metyrapone, even in normal individuals.

Dexamethasone suppression test

Another method of testing the feedback control is by determining the suppressive effect of supra-physiological doses of a glucocorticoid on ACTH release. Dexamethasone in a dosage of 0·5 mg 6-hourly for 2 days does not interfere with steroid estimations and the level of urinary 17-oxogenic steroids normally falls to less than 5 mg per 24 hours at the end of the test.

A rapid suppression test can be carried out conveniently in combination with the determination of the diurnal variation in plasma cortisol. Blood is taken at 9 a.m. for estimation of plasma cortisol and again at midnight at which time 2 mg of dexamethasone is given orally in one dose. Blood is taken again at 9 a.m. the following morning. The plasma cortisol level normally falls to less than 5 µg per 100 ml.

The dexamethasone suppression test is most often used in the diagnosis of Cushing's syndrome, and if larger doses of 8 mg daily are given it also provides information about the pathogenesis of the disease (see Chapter 9). Failure to react normally to the dexamethasone suppression test also occurs in some lesions of the central nervous system.

The pyrogen test

It is important to be able to assess the ability of a patient, whose pituitary function is in doubt, to react to stressful stimuli. Several procedures have been devised, one of which depends on the increase in plasma corticosteroids which follows the administration of a bacterial pyrogen. The writer has used a lipo-polysaccharide endotoxin derived from a species of *proteus* (Organon), given intravenously in the dosage of 0·005 µg per kg body weight. Blood is taken for plasma corticosteroid estimation before the injection and three hours later. The normal mean increase in plasma cortisol is 22·1 ± 10.1 (S.D.) µg per 100 ml. with a range of 8·6 to 45·0 µg (Jenkins, 1968). The increase is not, however, dependent upon the febrile reaction, which is very variable in degree and may not coincide with the maximal steroid response.

Side effects. Muscle pain and headaches are usually slight, and can be greatly reduced by the use of aspirin or similar analgesic, since this does not interfere with the steroid response. The procedure is terminated by the intravenous injection of 100 mg of cortisol succinate, and this should always be at hand. Providing these measures are taken and the dosage recommended is not exceeded, the test can be performed quite safely with relatively little discomfort. Like other workers, the writer has found many patients with pituitary lesions who react subnormally to metyrapone but normally to pyrogen, indicating that different sites of pituitary adrenal control are assessed by the two procedures.

Insulin hypoglycaemia test

The stressful stimulus most often used other than pyrogen is hypoglycaemia induced by an intravenous injection of insulin. Landon *et al.* (1963) recommend an initial dose of 0·1 unit of insulin per kg body weight in patients suspected of hypopituitarism. Blood is taken before the injection, and at 30, 60, and 90 minutes later for estimation of blood sugar and cortisol. It is essential that the blood sugar falls to at least 40 mg per 100 ml. in order to obtain a satisfactory response, and it may therefore be necessary to increase the insulin dosage. The maximal mean increase in cortisol is 12·8 ± 3·7 (S.D.) μg per 100 ml. with a range of 5·4–18·4 μg. The risk of producing profound hypoglycaemia in a hypopituitary subject requires that the patient is kept under close observation and that glucose and cortisol for intravenous use are readily available.

Vasopressin test

It has been known for many years that vasopressin, if given in sufficiently high dosage, promotes the release of ACTH from the anterior pituitary of experimental animals. If the dose is very high there is also some evidence of a direct ACTH-like effect on the adrenal cortex. It is now clear that vasopressin is not the physiological ACTH-releasing factor produced by the hypothalmus, but a test of pituitary-adrenal function based upon its use has been devised. The synthetic octapeptide, lysine vasopressin (Sandoz) is used, and is most often given intramuscularly in a dosage of 10 units. Blood is taken for the estimation of cortisol before the injection and again one hour later when the effect is maximal. The normal mean increase in plasma cortisol is 17·4 ± 6·8 (S.D.) per μg per 100 ml. with a range of 8·3 to 32·0 μg (Jenkins, 1968).

Side effects. Skin pallor is usual and occasionally abdominal colic is experienced. No significant increase in blood pressure has been noted, but caution is advised in the use of this test, since vasopressin has been reported to induce coronary artery constriction. Patients with a history of coronary artery disease should be excluded.

The choice of pituitary-adrenal function tests

Pituitary-adrenal function tests are used in the investigation of lesions of the hypothalamus and anterior pituitary. They can also be of great value in the assessment of the hypothalmic-pituitary-adrenal axis in patients who have received long-continued adrenal steroid therapy. An ACTH test is necessary in order to demonstrate that the adrenal cortex is responsive. In the view of the writer the metyrapone test should be retained in spite of occasional difficulties in interpretation of the normal response, because it is a very sensitive indicator of pituitary damage. A normal response probably excludes any significant degree of pituitary insufficiency. A subnormal response to metyrapone, however, does not give any indication of the reaction to a stressful stimulus, which in many ways is of more practical importance. For this reason a pyrogen, insulin, or vasopressin test should also be used. Although they are often termed "stress tests" the exact site of action of these agents has not yet been completely defined. It seems very likely that the action of vasopressin, in the doses used, lies at the level of the hypothalamus, whereas there is some evidence that pyrogen may act directly on the pituitary (Jenkins, 1968).

When considering the choice, it is the opinion of the writer that at the present time the pyrogen or insulin test should be used, and that more experience of the vasopressin test and its safety is desirable before recommending its widespread adoption.

Information on various aspects of pituitary-adrenal function tests and their limitations is to be found in "The assessment of hypothalmic-pituitary-adrenal function", *Memoirs of the Society for Endocrinology* No. 17, edited by James & Landon (1968).

REFERENCES

JAMES, V. H. T. and LANDON, J. (1968). Editors of "The Assessment of hypothalamic-pituitary-adrenal function", *Memoirs of the Society for Endocrinology* No. 17, Cambridge University Press.

JENKINS, J. S. (1968). In *Memoirs of the Society for Endocrinology*, No. 17, pp. 205–212. Ed. by V. H. T. JAMES and J. LANDON. Cambridge University Press.

JENKINS, J. S. and ELKINGTON, S. G. (1964). *Lancet*, **2**, 991–994.

LANDON, J., WYNN, V., and JAMES, V. H. T. (1963). *J. Endocr.* **27**, 183–192.

LIDDLE, G. W., ESTEP, H. L., KENDALL, J. W., WILLIAMS, W. C., and TOWNES, A. W. (1959). *J. clin. Endocr.* **19**, 875–894.

WOOD, J. B., FRANKLAND, A. W., JAMES, V. H. T., and LANDON, J. (1965). *Lancet*, **1**, 243–245.

13

CHROMATOGRAPHY OF ADRENAL STEROIDS

The procedures for the estimation of adrenal steroids in blood and urine described in previous chapters measure classes of compounds which have a common side group. The estimation of a specific steroid requires that it should first be isolated from the complex mixture in which it exists in body fluids. This separation is most often carried out by chromatography, and without this technique none of the modern knowledge of adrenal steroid secretion would have been obtained. All forms of chromatography depend in principle upon a distribution of the substance to be analysed between two phases. Four main types are in use at the present time (1) column (2) paper (3) thin-layer (4) gas–liquid chromatography (Table 13.1).

(1) COLUMN CHROMATOGRAPHY

This technique was used in the classical separation of adrenal steroids by Reichstein, Kendall, and others in the 1930s. The procedure is still much used, and is usually divided into adsorption and partition methods. In practice, however, the distinction is not complete since adsorption plays some part in most partition systems.

A glass tube is packed with a supporting material consisting usually of powdered alumina, magnesium silicate, or silica (silica gel). In adsorption chromatography the steroid mixture is dissolved in a solvent in which it is just soluble and is applied to the column. The alumina or silica adsorbs the steroids which are then removed by allowing a series of suitable solvents to flow down the column, the effect being a kind of partition between the solid support and the eluting solvent. In what is called "partition column chromatogaphy", two liquid phases are used, one of which is termed the stationary phase and attaches itself to the support, and the other is the mobile phase. The mixture of steroids is applied to the top of the column and is carried down by the mobile phase. Separation of the individual constituents

TABLE 13.1. Chromatographic methods

	Column	Paper	Thin-layer	Gas-liquid
Main uses	Preparatory separation. 17-oxosteroids Pregnanetriol.	Separation of all adrenal steroids.	Separation of all adrenal steroids.	17-oxosteroids Pregnanetriol. Corticosteroids only after oxidation. Also used for pregnanediol, oestrogens, and testosterone.
Advantages	Large capacity. Can handle "impure" extracts.	Good separation of closely related compounds.	Rapid. Compact spots	Rapid, Good separation with simultaneous quantitation. Very high sensitivity.
Disadvantages	Separation cannot be directly seen. Alumina destroys corticosteroids.	Fairly slow. Requires extracts to be free from lipid for good resolution.	Requires specially prepared plates. Separation of very closely related compounds not easily achieved.	Expensive apparatus. Thermal instability of some steroids. Extensive preliminary purification required.

takes place according to the differences in partition between the two liquid phases. Examples of the solvent systems used are shown in Table 13.2. Column chromatography can accomodate wide ranges in amounts of material and is therefore used in preparative work. It is much used as a preliminary purifying procedure before carrying out further chromatographic separation on paper or gas-liquid systems. It is also used in the simple separation of 17-oxosteroids into 11-oxy and 11-deoxy groups, and for the routine estimation of pregnanetriol.

(2) PAPER CHROMATOGRAPHY

The use of paper as a supporting medium for a partition type of chromatography was introduced originally for the separation of water-soluble compounds. The application of the technique to steroid separations was described by Zaffaroni in 1950 and Bush in 1952, and their procedures have been largely responsible for the great progress in steroid research which has taken place in recent years. By the use of simple, inexpensive apparatus which is within the province of most laboratories, separation of steroids can be carried out on the microgram scale. In the Zaffaroni method, sheets or strips of paper are impregnated with a relatively non-volatile stationary phase such as propylene glycol, the steroid mixture is applied in a spot or line close to one end and the paper is allowed to hang from a trough suspended in a closed tank. The chromatogram is developed by filling the trough with mobile phase such as toluene saturated with propylene glycol, which flows down the paper, and the separation of steroids occurs by partition between the two phases. In Bush's systems solvents are more volatile, the stationary phase being aqueous methanol. The steroid mixture is first applied to the paper which is then hung from a trough in a sealed tank containing both phases for a minimum of 3 hours. The paper adsorbs the stationary phase, and at the end of the period of equilibration, mobile phase is then added to the trough from which it starts to flow down the paper. It reaches the end of a 45 cm strip in about 3–4 hours at room temperature, but can be allowed to run as long as is desirable to effect good separation.

Although the paper methods involve partition between mobile and stationary phases, it is likely that adsorption on the paper also plays a part in the separation obtained. For a discussion of the complex factors upon which chromatographic systems depend, the monograph by Bush (1961) should be consulted.

Detection of steroids on the paper
Several procedures have been devised to localize the steroid spots on the chromatogram, most of which are sensitive to 5 μg of material or less. The most commonly used methods of detection are shown in Table 13.3.

TABLE 13.2. Examples of commonly used solvent systems for chromatography of adrenal steroids

Column	Adsorption	Elution from alumina with petroleum ether-benzene or various concentrations of benzene-ethanol.
		Elution from silica gel with benzene-ethylacetate.
	Partition	Kieselguhr column; benzene as mobile phase and aqueous methanol as stationary phase.
Paper	Zaffaroni system	Toluene saturated with propylene glycol as mobile phase. Propylene glycol as stationary phase.
	Bush system	Benzene, toluene or petroleum ether saturated with aqueous methanol as mobile phase. Aqueous methanol as stationary phase.
Thin-layer		Various mixtures, e.g. chloroform and ethanol or benzene and ethyl acetate.
Gas-liquid		Argon or nitrogen as mobile phase. Various silicones as stationary phase.

TABLE 13.3. Commonly used methods of detection of steroids on paper and thin-layer plates.

Method	Steroids detected	Approximate limit of sensitivity (μg)
Absorption of U.V. light at 254 mμ (paper only)	Δ⁴-3-oxo steroids	5
Absorption of U.V. light fluorescence at 254 mμ (Special thin-layer plates only)	Δ⁴-3-oxo steroids	0·2
Reduction of blue-tetrazolium to blue colour	α-ketols	3—5
Yellow fluorescence under light of 365 mμ when treated with NaOH and dried. (Paper only)	Δ⁴-3-oxo steroids	0·25—1
Alkaline m-dinitro benzene. Concentrated sulphuric acid gives various colours followed by charring (Thin-layer only)	17-oxosteroids All steroids	3—5 0·1

The running rate characteristic of a steroid in a particular system is denoted by its R_F value which is

$$\frac{\text{distance from origin to centre of steroid spot}}{\text{distance from origin to solvent front}}$$

If the solvent has run off the paper the distance travelled by a steroid can be expressed in relation to that of a reference steroid such as cortisol. Standard known steroids are run in parallel with the mixture to be analysed, and if the R_F value of the unknown is similar to that of the standard, its possible identity is suggested. It is very important, however, to realize that similar running characteristics in a single solvent system afford *no proof* of identity. In order to establish this, it is necessary to show that the R_F value of the unknown is similar to the authentic compound in several different systems, and, more important, to show that the product obtained after chemical reaction such as acetylation and oxidation also behaves similarly to that of the standard treated in a similar fashion.

Quantitation of steroids after paper chromatography can be carried out directly on the paper by applying one of the colour reactions used in detection, followed by some form of densitometry or fluorimetry. When greater accuracy is required the compounds are eluted from the paper and then such procedures as ultra-violet light absorption, Porter-Silber or Zimmerman reactions are used, according to the particular side-groups of the steroid to be analysed. In the most precise work traces of radioactively labelled compounds are added at the beginning of the extraction to enable a correction for losses to be made.

(3) THIN-LAYER CHROMATOGRAPHY

This is a more recent addition to the available chromatographic procedures for separation of steroids. Glass plates are coated with a thin uniform layer of silica gel or alumina and the steroid mixture is applied 2 cm from one margin with standards in parallel. This edge of the plate is then placed on the bottom of a sealed tank containing a solvent mixture to a depth of about 1 cm. The solvent ascends the plate by capillary action and reaches the upper edge in about 30–60 minutes. The plate is then removed, dried, and the spots can be located by methods shown in Table 13.3. A useful method of detection employs a fluorescent material incorporated in the silica gel, which allows Δ^4-3-oxo compounds to be seen as dark spots against a fluorescent background under light of wavelength 254 mμ. The advantage of thin-layer chromatography is that it is much more rapidly completed than the paper methods and it gives more compact spots. It requires, however, specially prepared plates and for separations of very closely similar compounds paper chromatography is probably superior.

(4) GAS-LIQUID CHROMATOGRAPHY

The use of gas–liquid chromatography for steroid separations is relatively recent and is still being developed. The principle is similar to liquid–liquid column chromatography except that gas takes the place of a mobile phase. An inert carrier gas such as argon or nitrogen flows through a long column, usually several feet in length, containing a support made of a diatomaceous earth upon which has been deposited the stationary phase. This phase is most often a silicone. The column is maintained at a high temperature, often between 190° and 250°C, so that when the steroid mixture is injected into one end it is immediately vaporized. Separation of the steroids occurs during their passage along the column by the process of partition between gas and stationary phase. As the steroids arrive at the distal end of the column they are detected and simultaneously measured by a device known as an ionization detector. In the most modern instruments the detector utilizes the ions formed during the combustion of hydrogen and when the steroid vapour mixed with carrier gas reaches the hydrogen flame an increase in ionization occurs, which, after suitable amplification, is recorded electrically as a peak. The area of a peak provides a measure of the amount of steroid. The volume of gas which flows before a particular steroid peak appears is known as the retention volume and the time which elapses before the height of the peak is seen is the retention time. In gas-liquid chromatography the retention-time is a useful characteristic of a steroid and is analogous to the R_F value of paper chromatography. In order to obtain better resolution and greater thermal stability steroids are usually converted into derivatives before they are chromatographed. The acetates and trimethyl silyl ethers are at present the most useful. Even so, those steroids with an α-ketol side chain, which include cortisol, corticosterone, aldosterone, and their metabolites, undergo marked decomposition at the elevated temperatures which are necessary, and at present they cannot be chromatographed satisfactorily except after preliminary oxidation. This has the disadvantage of reducing specificity since several corticosteroids may yield a common 17-oxosteroid on oxidation. Aldosterone, however, can be chromatographed after formation of its γ-lactone by oxidation with periodic acid (Bravo & Travis, 1967). The advantages of gas-liquid chromatography are good separation with simultaneous quantitation, and extreme sensitivity. If a special form of detector of the electron-capture type is used it is possible to carry out estimations within the range of $10^{-5}\mu g$, and even with the more conventional flame ionization detector, the measurement of $10^{-2}\mu g$ can be easily achieved.

H

Considerable skill is required, however, in order to exploit the gas-chromato-graph at these very high degrees of sensitivity. Disadvantages are the expense of the apparatus, the need in most instances for extensive preliminary puri-fication of steroid extracts followed by the formation of derivatives before they are suitable for application to the gas chromatograph, and the thermal instability which limits the range of steroids which can be chromatographed. Further developments in this field are, however, taking place rapidly and the final place of gas-liquid chromatography in steroid methodology cannot yet be foreseen.

This survey of chromatographic methods is of necessity brief, but a full exposition of the subject is to be found in the excellent monographs by Bush (1961) and by Neher (1964). Useful sources of information on gas–liquid chromatography of steroids are provided by Wotiz & Clark (1966) and Grant (1967).

REFERENCES

Bravo, E. L. and Travis, R. H. (1967). *J. Lab. clin. Med.* **10**, 831–840.
Bush, I. E. (1961). *The Chromatography of Steroids*. Pergamon Press, Oxford.
Grant, J. K. (1967). Editor of *"The Gas Liquid Chromatography of Steroids"*. *Memoirs of the Society for Endocrinology* No 16, Cambridge University Press.
Neher, B. (1964) *Steroid Chromatography*. Elsevier, Amsterdam.
Wotiz, H. H. and Clark, S. J. (1966). *Gas Chromatography in the Analysis of Steroid Hormones*. Plenum Press, New York.

14

ADRENAL STEROID SECRETION RATES

The estimation of the secretion rate of adrenal steroid hormones, particularly that of cortisol, provides a direct method for the assessment of adrenocortical function, but facilities for counting radioactive isotopes and for paper chromatography are required. The technique depends upon the principle of isotope dilution in which an administered dose of radioactive steroid becomes diluted by the endogenous secretion of the steroid. The dilution can be determined either in the blood by taking serial blood samples, or in the urine over a period of 24 hours. The urine method is the simpler and the first step is the separation from urine of a "unique" metabolite of the steroid, i.e. a metabolite which is derived exclusively from the steroid to be investigated. For the determination of cortisol secretion rates tetrahydrocortisol or tetrahydrocortisone are unique metabolites, and for aldosterone the acid-labile conjugate or tetrahydroaldosterone can be used.

THE ESTIMATION OF THE CORTISOL SECRETION RATE

A known dose of cortisol radioactively labelled with carbon-14 or tritium is administered to the patient either intravenously or orally. Usually a dose of about 1 μc is used. Cope & Pearson (1965) have shown that in spite of theoretical objections to the oral administration there is ordinarily little difference in the results obtained with either route. Following the administration of the labelled cortisol, urine is collected for 24 hours, and for a second 24 hours if there is any impairment of renal function. The glucuronide conjugates in an aliquot of the 24-hour sample are hydrolised with β-glucuronidase, and tetrahydrocortisol or tetrahydrocortisone is isolated by paper chromatography. The metabolite is eluted from the paper, the radio-activity is counted and the amount of steroid present is determined using the Porter-Silber or blue-tetrazolium reaction. The specific activity of the metabolite (radioactivity in counts per minute per mg) is then calculated. It should be noted that exact quantitation is only required for the determination of the specific activity and that any losses of material encountered in the isolation of the metabolite are of no account. If the urine is collected for 24 hours, the secretion rate in mg for this period is then given by:

$$\frac{\text{dose administered (counts per min)}}{\text{specific activity of metabolite (counts per min per mg)}}$$

This formula assumes that 100% of the radioactivity administered is excreted in the urine within 24 hours. Ordinarily about 75–90% is excreted but for clinical purposes the error involved in using the above formula is small providing renal function is normal.

The determination of secretion rates by this technique requries that several conditions should be fulfilled: isotopically labelled steroid must be metabolized in the same manner as the endogenous steroid; complete mixing of isotopically labelled and endogenous steroids must take place; the metabolite selected for estimation of the specific activity must be truly "unique"; the isotopic label must be stable; excretion of the radioactive metabolite should be almost complete; the weight of isotopically labelled steroid which is administered should be negligible compared with the amount of andogenous steroid secreted.

If renal function is impaired there may be a considerable delay in the excretion of radioactivity, and considerable error is then introduced. Cope & Pearson (1963) have attempted to correct for this by estimating the total radioactivity in the urine both in the first and the second 24 hours after administration of the labelled steroid.

Their corrected formula for the secretion rate is then:

$$\frac{\text{dose administered}}{\text{specific activity of metabolite}} \times \left(1 - \frac{B}{U}\right)$$

where U is percentage of isotope excreted in urine in the first 24 hours and B the percentage in the second 24 hours. It seems likely, however, that with increasing degrees of renal failure the estimation becomes very approximate.

The use of the cortisol secretion rate

The normal range in the adult is given by Cope & Pearson (1965) as 6·3 to 28·6 mg daily with a mean figure of 16·2 ± 5·7 S.D. These values are in accordance with the amount of cortisol which is required to maintain the health of a patient with Addison's disease, and this is usually about 20 mg cortisol daily or rather more of cortisone. In infants the cortisol secretion rate is about 3 mg per 24 hours and increases with age, but is equivalent to that of the adult when expressed on the basis of surface area, the normal secretion rate being $12·0 ± 2·0$ mg/m^2 per 24 hours. Moderately raised values are found in obesity (Migeon et al. 1963). In late pregnancy the cortisol secretion rate is about $2\frac{1}{2}$ times that of the non-pregnant woman.

Increased secretion rates are also found in thyrotoxicosis, and conversely they are decreased in myxoedema and in hepatic cirrhosis. These results are in accordance with the increase and decrease in metabolism of cortisol which occurs in these conditions so that the levels in the plasma remain within the normal range.

The main use of the cortisol secretion rate is in the diagnosis of those cases of Cushing's syndrome where urinary 17-oxogenic steroid excretion or plasma corticosteroid levels are within the normal range. Cope and Pearson report that in Cushing's syndrome the cortisol secretion rate ranges from 36 to 138 mg daily in cases of adrenal hyperplasia, and in adrenal carcinoma figures of 137, 211, and 316 mg were obtained. Cope has shown that when the cortisol secretion rate exceeds 35 mg daily the 17-oxogenic steroid excretion is about 50% of the secretion, but this relationship does not always hold, and in some cases of Cushing's syndrome with normal 17-oxogenic steroid values the cortisol secretion rate was found to be between 35 and 60 mg. In the near normal to low range of secretion, there is no regular correlation with urinary steroid excretion.

In severe *hypopituitarism* very low cortisol secretion rates are found but in Addison's disease the range in the series reported by Cope and Pearson was 0·6 to 19 mg per day, demonstrating that the adrenal lesion is sometimes incomplete.

Corticosterone

The secretion rates of corticosterone are not ordinarily of much clinical value in man but Peterson & Pierce (1960) have reported normal values ranging from 1·5 to 4·0 mg daily with a mean of 2·3 mg. The estimation of aldosterone secretion rates is discussed in Chapter 15.

Cope & Pearson (1965) give a good survey of the clinical use of the cortisol secretion rate, and Tait (1963) discusses the theoretical basis of various methods for the estimation of secretion rates.

REFERENCES

COPE, C. L. and PEARSON, J. (1963). *Clin. Sci.* **25**, 331–341.
COPE, C. L. and PEARSON, J. (1965). *J. clin. Path.* **18**, 82–87.
MIGEON, C. J., GREEN, O. C., and ECKERT, J. P. (1963). *Metabolism*, **12**, 718–739.
PETERSON, R. E. and PIERCE, C. E. (1960). *J. clin. Invest.* **39**, 741–757.
TAIT, J. F. (1963). *J. clin. Endocr.* **23**, 1285–1297.

15

THE ESTIMATION OF ALDOSTERONE

Particular difficulties are associated with the estimation of aldosterone owing to the very small amounts which are present in body fluids.

ESTIMATION OF ALDOSTERONE IN URINE

The two main metabolites of aldosterone are the tetrahydro-derivative and the acid-labile conjugate, but the latter is most often measured. Although this metabolite normally constitutes only about 10% of the total aldosterone secretion it can be easily separated, after acid hydrolysis, from the large amount of cortisol metabolites which remains in the urine as glucuronides conjugated at ring A. Some workers, however, estimate tetrahydroaldosterone and in this case hydrolysis with β-glucuronidase is first necessary.

The first practicable method for the estimation of the acid-labile conjugate was described by Neher & Wettstein (1956). In this procedure half the volume of a 24-hour specimen of urine is brought to pH1 and allowed to stand for 24 hours at room temperature. Extraction with chloroform is then carried out followed by a wash with sodium hydroxide, and evaporation to dryness. The extract is then subjected to column chromatography, followed by two different paper chromatographic separations. The aldosterone is located by the blue tetrazolium reaction and estimated visually by the yellow fluorescence seen in ultra-violet light after treating with aqueous sodium hydroxide and drying the paper. Normal values of 1–9 µg daily were obtained by Neher and Wettstein. In spite of three chromatographic separations the method has been criticized on grounds of lack of specificity and is now very little used in its original form. Modifications have been introduced in order to improve the specificity, particularly acetylation of the aldosterone, followed by a fourth chromatogram, but losses are often considerable.

The double isotope technique

The best method is that devised by Kliman & Peterson (1960) and depends upon the simultaneous use of two radioactive isotopes. In this technique one isotope is used for the actual estimation of the steroid and the second isotope is introduced near the beginning of the procedure to enable a correction to be made for the losses which occur.

A small aliquot from a 24-hour sample of urine is allowed to stand at pH1 for 24 hours and extracted with dichloromethane, following which the extract is acetylated with acetic anhydride labelled with tritium. A known but very small quantity of aldosterone diacetate labelled with carbon-14 is then added to serve as a marker. The total aldosterone diacetate in the extract is then separated by two different paper chromatographic systems, eluted from the paper and oxidized with chromic acid. The product is subjected to further paper chromatography, eluted from the paper, and the radioactivity is counted in a liquid scintillation spectrometer capable of counting tritium and carbon-14 separately. No chemical reaction is therefore involved in the actual estimation of aldosterone, which is determined by the tritium counts. The carbon-14 counts serve to correct for losses, which may be 75% or more. The method is extremely sensitive and it is possible to estimate the aldosterone in aliquots of urine containing less than 0·1 µg.

Normal values for subjects taking an ordinary diet are 5–19 µg daily. Levels are increased when normal subjects are depleted of sodium, and during the later stages of pregnancy. They are higher when the subject is in the upright position compared with the recumbent posture. Pathologically, raised values are found in some patients suffering from nephrosis, cirrhosis of the liver with ascites, and malignant hypertension. In congestive heart failure the levels are sometimes raised but often normal.

It is obvious that this technique is in no sense a routine procedure since it is extremely time-consuming, considerable technical skill is necessary and expensive radioactive counting equipment is required.

The clinical use of aldosterone estimations

Apart from research into aldosterone secretion the estimation of aldosterone levels in urine is most often sought in the diagnosis of primary hyperaldosteronism (Conn's syndrome). Before a request is made for such a laborious estimation, however, it is essential that the diagnosis should have already reached a high degree of certainty from consideration of the simpler biochemical changes outlined in Chapter 9. Moreover, the estimation may not always be helpful since some cases of Conn's syndrome have urinary aldosterone levels which are within the normal range, and in such cases estimation of the aldosterone secretion rate may be more useful.

ESTIMATION OF THE ALDOSTERONE SECRETION RATE

This procedure is carried out by the isotope dilution method in a manner similar to that described in Chapter 14 for the cortisol secretion rate although the estimation is technically much more difficult. The specific activity of a urinary metabolite of aldosterone is determined after the intravenous administration of a known dose of d-aldosterone labelled with tritium. The "unique" metabolite which is used can be either the acid-labile conjugate or tetrahydroaldosterone. The latter metabolite has been used by Cope et al. (1961) who estimated tetrahydroaldosterone in urine by acetylation with acetic anhydride labelled with carbon-14, the separation being achieved by three chromatographic steps. The secretion rate is then calculated from the dose of tritium-labelled aldosterone administered divided by the tritium content per microgram of the isolated metabolite. Cope et al. reported that the aldosterone secretion rate in normal subjects on a normal diet ranged from 62 to 275 µg daily with a mean value of 144 µg. Other workers using the acid-labile conjugate have found a similar order of values (Mills, 1962). Subjects given a low sodium diet had secretion rates which ranged from 880 to 1817 µg daily. In normal pregnancy, especially in the third trimester, values may reach about ten times that of the non-pregnant state. Of interest is the finding that in cases of severe toxaemia of pregnancy the secretion rate is not elevated. In hepatic cirrhosis with ascites both high and low values have been reported and in congestive heart failure values are occasionally raised but are more often within the normal range. Very high aldosterone secretion rates of up to 6,600 µg daily have been recorded in patients with nephrosis, and extremely high figures have also been found in some cases of malignant hypertension. Some of these patients may have varying degrees of renal failure, however, and under these conditions falsely high figures can be obtained due to incomplete urinary excretion of the aldosterone metabolites. The correction for errors in steroid secretion rates in renal failure by the formula devised by Cope and Pearson is described in Chapter 14.

In primary hyperaldosteronism the aldosterone secretion rate is elevated but often not to the same degree as in some cases of secondary hyperaldosteronism, so that values greater than 1,000 µg daily are unusual. Abnormally low values which do not rise following sodium restriction are found in infants with congenital adrenal hyperplasia of the salt-losing type. Cases which do not lose salt have normal secretion rates. Normal infants have aldosterone secretion rates of 22–54 µg daily, in a series aged from three days to nine months, reported by Kowarski et al. (1965). These authors report a range for normal adults which is somewhat lower than that found by other workers so that the values for infants are comparatively high.

The estimation of aldosterone in blood

The concentration of aldosterone in the peripheral blood is rather less than one-thousandth that of cortisol, so that the estimation is associated with formidable difficulties. A double isotope technique is essential but in the method devised by Peterson, based upon his method for urinary aldosterone estimation, ^{14}C-acetic anhydride of very high specific activity is required, so that on grounds of cost alone it has been performed in very few centres. Brodie *et al.* (1967), however, have reduced the cost considerably by preparing the tritium-labelled aldosterone monoacetate instead of the ^{14}C-labelled diacetate used by Peterson. These workers report plasma aldosterone levels of 1·25–13·6 mμg per 100 ml. in normal individuals on a normal diet with a mean of 6·97 mμg and these figures are closely similar to those of Peterson. Such a procedure is used for research purposes only and is of value in studying rapid changes in aldosterone secretion. It has been shown that posture is an important factor in determining the level, higher values being found when subjects are upright than when they are recumbent. This explains why aldosterone secretion is apparently higher during the day than at night, but there does not appear to be a diurnal variation comparable to that which is present in cortisol levels. Plasma aldosterone concentration in normal subjects is increased by sodium depletion and the infusion of angiotensin. The administration of ACTH in physiological amounts also causes a rise in levels, but the increase is only transient (James & Fraser, 1966).

REFERENCES

BRODIE, A. H., SHIMIZU, N., TAIT, S. A. S., and TAIT, J. F. (1967). *J. clin. Endocr.* **27,** 997–1011.
COPE, C. L., NICOLIS, G., and FRASER, B. (1961). *Clin. Sci.* **21,** 367–380.
JAMES, V. H. T. and FRASER, R. (1966). *J. Endocr.* **34,** xvi–xvii.
KLIMAN, B. and PETERSON, R. E. (1960). *J. Biol. Chem.* **235,** 1639–1648.
KOWARSKI, A., FINKELSTEIN, J. W., SPAULDING, J. S., HOLMAN, G. H., and MIGEON, C. J (1965). *J. clin. Invest.* **44,** 1505–1513.
MILLS, J. N. (1962). *Brit. Med. Bull.* **18,** 170–173.
NEHER, R. and WETTSTEIN, A. (1956). *J. clin. Invest.* **35,** 800–805.

INDEX